ANOMIA

FOUNDATIONS
OF
NEUROPSYCHOLOGY

A Series of Textbooks, Monographs, and Treatises

Series Editor

LAIRD S. CERMAK

*Memory Disorders Research Center, Boston Veterans Administration,
Medical Center, Boston, Massachusetts*

ANOMIA
Neuroanatomical and Cognitive Correlates

Edited by

Harold Goodglass

Aphasia Research Center
Department of Neurology
Boston University School of Medicine
VA Medical Center
Boston, Massachusetts

Arthur Wingfield

Department of Psychology, and
Volen National Center for Complex Systems
Brandeis University
Waltham, Massachusetts

Academic Press

San Diego   London   Boston   New York   Sydney   Tokyo   Toronto

Copyright © 1997 by ACADEMIC PRESS

Academic Press
a division of Harcourt Brace & Company
525 B Street, Suite 1900, San Diego, California 92101-4495, USA
http://www.apnet.com

Academic Press Limited
24-28 Oval Road, London NW1 7DX, UK
http://www.hbuk.co.uk/ap/

Library of Congress Cataloging-in-Publication Data

Anomia : neuroanatomical and cognitive correlates / edited by Harold
    Goodglass, Arthur Wingfield.
        p.   cm. -- (Foundations of neuropsychology)
    Includes index.
    ISBN 0-12-289685-8 (alk. paper)
    1. Anomia.   I. Goodglass, Harold.   II. Wingfield, Arthur.
    III. Series: Foundations of neuropsychology (San Diego, Calif.)
    [DNLM: 1. Anomia.   WL 340.5 A615 1997]
    RC425.6.A56   1997
    616.85'52--dc21
    DNLM/DLC
    for Library of Congress   96-49185
                                                               CIP

PRINTED IN THE UNITED STATES OF AMERICA
97 98 99 00 01 02 EB 9 8 7 6 5 4 3 2 1

Contents

Chapter 7   *Naming in Normal Aging and Dementia of the Alzheimer's Type*

Marjorie Nicholas, Christine Barth, Loraine K. Obler, Rhoda Au, and Martin L. Albert

Chapter 8   *Treatment of Aphasic Naming Problems*

Nancy Helm-Estabrooks

Chapter 9   *Summary of the Volume*

Harold Goodglass and Arthur Wingfield

Contributors

Numbers in parentheses indicate the pages on which the authors' contributions begin.

Martin L. Albert (165) Medical Research Service, Department of Veterans Affairs Medical Center, Boston, Massachusetts and Aphasia Research Center, Department of Neurology, Boston University School of Medicine, Boston, Massachusetts 02130

Rhoda Au (165) Department of Neurology, Boston University School of Medicine, Boston, Massachusetts 02118

Christine Barth (165) Psychology Department, University of Connecticut, Storrs, Connecticut 06320

Hanna Damasio (65) Department of Neurology, Division of Cognitive Neuroscience, University of Iowa College of Medicine, Iowa City, Iowa 52242, and The Salk Institute, La Jolla, California 92037

Antonio R. Damasio (65) Department of Neurology, Division of Cognitive Neuroscience, University of Iowa College of Medicine, Iowa City, Iowa 52242, and The Salk Institute, La Jolla, California 92037

R. De Bleser (93) Patholinguistics/Cognitive Neurolinguistics, Institute of Linguistics, University of Potsdam, D-14415 Potsdam, Germany

Harold Goodglass (3, 203) Aphasia Research Center, Department of Neurology, Boston University School of Medicine, VA Medical Center, Boston, Massachusetts 02130

Barry Gordon (31) Departments of Neurology and Cognitive Science, and the Zanvyl Krieger Mind/Brain Institute, The Johns Hopkins University, Baltimore, Maryland 21205

Nancy Helm-Estabrooks (191) Department of Neurology, Boston University School of Medicine, Boston, Massachusetts 02130, and National Center for Neurogenic Communication Disorders, University of Arizona, Tucson, Arizona 85121

Paula Menyuk (137) Developmental Studies, Boston University, Boston, Massachusetts 02215

Marjorie Nicholas (165) Medical Research Service, Department of Veterans Affairs Medical Center, Boston, Massachusetts 02130

Loraine K. Obler (165) City University of New York Graduate School, New York, New York 10021, and Medical Research Service, Department of Veterans Affairs Medical Center, Boston, Massachusetts 02130

Carlo Semenza (115) Department of Psychology, University of Padova, Padova, Italy

Daniel Tranel (65) Department of Neurology, Division of Cognitive Neuroscience, University of Iowa College of Medicine, Iowa City, Iowa 52242

Arthur Wingfield (3, 203) Department of Psychology, and Volen National Center for Complex Systems, Brandeis University, Waltham, Massachusetts 02254

Preface

The emergence of the ability to name objects is among the earliest accomplishments in the development of language in the young child. During the second through the fourth years of life, new words are acquired at the rate of two to four a day. Although the rate and extent of vocabulary growth are known to be correlated with intellectual ability, the capacity to refer to objects or concepts by name appears as a fundamental capacity for every neurologically intact child well before school age.

On the other side of this coin, the impairment of this ability is an almost constant finding in individuals with structural injury to the language zone of the left hemisphere, with resulting aphasia. *Anomia* refers to the inability to access spoken names for objects or other concepts in aphasic patients who would otherwise have sufficient articulatory facility to produce the words if they could be retrieved. When the ability to access object names is selectively impaired despite normal comprehension and fluent production of sentence forms, a patient is said to have *anomic aphasia*.

Our goal in this volume is to offer a state-of-the-art review of disorders of naming, approached from both clinical and theoretical viewpoints. The volume begins with an overview of naming and aphasia by the editors. The succeeding chapters were prepared by acknowledged experts who provide comprehensive literature reviews, summaries of relevant research data, and interpretive integrations of the work in each of their domains. This book is the first devoted entirely to naming and its disorders. Included are descriptions of advances in cognitive analysis and anatomic findings based on functional imaging and clinical observations that could not have been written 10 years ago. We believe this book will be useful to readers in the many disciplines involved in the study of language, including those in cognitive neuroscience, neurology, speech pathology, and linguistics.

We are grateful for the support and stimulation that we have gained from our many discussions of naming with our colleagues at the Boston University Aphasia Research Center and the Memory and Cognition Laboratory at Brandeis University. We acknowledge the support for our own research from National Institutes of Health Grants DC00081 and AG04517 and support from the W. M. Keck Foundation.

Harold Goodglass
Arthur Wingfield

Part I

Introduction

Word-Finding Deficits in Aphasia: Brain–Behavior Relations and Clinical Symptomatology

Harold Goodglass and Arthur Wingfield

Aphasia is a general term for a language impairment following brain damage. Such impairments can take the form of difficulty in fluent production of connected speech (e.g., Broca's aphasia), impaired speech comprehension (Wernicke's aphasia), a failure or difficulty in reading (alexia), writing (agraphia), repetition (conduction aphasia), or any combinations of these or other communicative syndromes.

Language deficits arise ordinarily from lesions in the left hemisphere in a region bounded anteriorly by the third frontal convolution, posteriorly to the angular gyrus in the parietal lobe, and vertically from the inferior temporal gyrus to the supramarginal gyrus. This area, generally surrounding the Sylvian fissure (the perisylvian region) has been properly called the language zone (see Figure 1). Lesions outside of this area rarely produce deficits of language, whereas damage within this area usually does.

Left-hemisphere laterality for language functions is by far the overwhelming rule in right-handers. Exceptions may occur, particularly in non-right-handers. An interesting coverage of anomalies of laterality in such cases is provided by Alexander, Fischette, and Fischer (1989).

The varieties of specific deficits that appear in aphasia are many, including difficulties related to the expression of syntax (agrammatism), music (amusia), calculation (acalculia), and so forth. Of the symptoms associated with aphasia, none are more pervasive than *anomia*, a difficulty in finding high information words, both in fluent discourse, and when called upon to identify an object or action by name.

*ANOMIA: Neuroanatomical
and Cognitive Correlates*

3

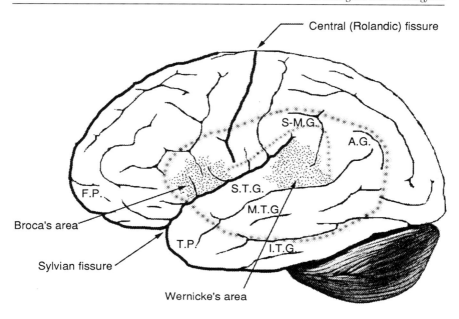

A.G.   Angular Gyrus
F.P.   Frontal Pole
I.T.G.   Inferior Temporal Gyrus
M.T.G.   Middle Temporal Gyrus
O.P.   Occipital Pole
S-M.G.   Supra-Marginal Gyrus
S.T.G.   Superior Temporal Gyrus
T.P.   Temporal Pole

Figure 1 *Above:* Lateral view of the left cerebral hemisphere showing the perisylvian language zone. (Adapted from Goodglass, 1993, Fig. 3.1., p. 40.) *Next Page:* Horizontal section of the brain at the level of Wernicke's and Broca's area. (Adapted from Goodglass, 1993, Fig. 3.2, p. 41.)

A common feature of anomia is that the words the patient is unable to produce are not lost from memory. A patient unable to produce the desired name of an object may readily identify the object by pointing to it when its name is spoken. He or she may recognize whether or not an offered name is the correct name of the object, or the patient may produce the correct name after being prompted with its initial sounds. Rather than being erased from the lexicon, it is more accurate to say that the name, or its correct phonology, is inaccessible to retrieval. This is a source of frustration for the patient, and its explanation has long represented an intellectual puzzle in cognitive neuropsychology.

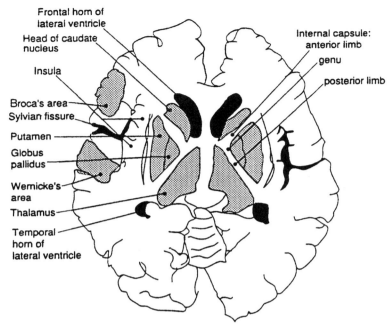

Frontal horn of
lateral ventricle

Head of caudate
nucleus

Insula

Broca's area

Sylvian fissure

Putamen

Globus
pallidus

Wernicke's
area

Thalamus

Temporal
horn of
lateral ventricle

Internal capsule:
anterior limb

genu

posterior limb

Figure 1 *Continued*

A QUESTION OF DEFINITION

As indicated above, we define anomia as the inability to retrieve the names for concepts that previously were readily available to the speaker. With good reason, this condition usually is identified with aphasia and hence with a lesion in the language zone of the brain. Indeed, virtually all individuals suffering from aphasia have some degree of word-retrieval problems, which may range from absolute failure, to a mild impairment. Patients who have largely recovered from aphasia following a brain lesion may have a persistent word-finding difficulty as their only residual effect.

Although the boundaries of the concept of anomia are not absolutely defined, it probably is not properly assigned to the sporadic difficulties in retrieving particular words and names that are a common complaint of older people (e.g., Burke, MacKay, Worthley, & Wade, 1991). These in turn are probably to be distinguished from the extensive loss of access to vocabulary observed in patients with Alzheimer's disease. The anomia of aphasia denotes a malfunction specific to the language system. By contrast, recent analytic studies of word-retrieval difficulties in patients with

Alzheimer's disease (Chertkow & Bub, 1990) relate the failures of these patients to extensive loss of semantic memory for the concepts they fail to name. The specifics of this research will be discussed in Chapter 7 by Barth et al., this volume, but they indicate that one probably can distinguish word-finding problems of Alzheimer's patients from those of aphasics.

The normal acquisition of naming in children, developmental abnormalities observed in the absence of focal injuries, and response to injury to the language zone after speech has been acquired all pose challenging questions to students of brain–language relationships. Because of brain plasticity in the early development of language, it is risky to attempt to apply an adult anatomo-functional template to understand childrens' disorders. Some of the clinical observations and theoretical accounts of developmental disorders of naming will be discussed in the Chapter 6 by Menyuk, this volume.

Although the term *anomia* may be used synonymously with word-finding difficulty, both of these terms have a restricted meaning: they refer to impaired retrieval of words that have a conceptual referent; they do not refer to the inability to produce the grammatical morphemes of the language, nor to the inability to produce articulate speech in general.

CLINICAL FORMS OF ANOMIA

Much of the clinician's sense that anomia takes many different forms arises from the accompanying deficits and adaptations available to patients with different forms of aphasia. Anomia may be in the foreground of the symptomatology or it may be overshadowed by other problems. The relative prominence of the anomia, as will be shown, is not necessarily tied to the absolute severity of the word-retrieval difficulty.

Broca's aphasics, whose most salient difficulty is in initiating and maintaining articulatory production, usually have a limited naming vocabulary. In free conversation, they rely predominantly on nouns to convey information in short, grammatically defective word groups.

Example of a profoundly agrammatic Broca's aphasic:
Examiner: What brought you into the hospital?
Patient: Yes, . . . ah . . . Monday . . . Dad . . . Paul H. (name deleted)—and dad . . . Hospital . . . Two . . . er . . . er . . . doctors . . . an' . . . thirty minutes . . . yes, . . . Hospital . . . and, er . . . We'sday, . . . We'sday, . . . nine o'clock . . . and, er . . . Thursday . . . ten o'clock . . . doctors . . . two . . . two . . . doctors.

Example of a moderately severe agrammatic Broca's aphasic:
Examiner: What work were you doing before you got sick?
Patient: Foreman.
E: Foreman?
Pt: Yeah.
E: Can you tell me what you had to do?
Pt: Su . . . pervise . . . supervise recor's . . . rectures . . . recor's . . . paper.
E: Records?
Pt: Yeah . . . ah . . . thass all . . . . Many kinds of work, but that's all.
E: How long had you been doing that?
Pt: One year . . . and planter two years . . . planner.
E: And what?
Pt: Two years . . . planner.
E: What did that consist of?
Pt: Time . . . motion studies.

Broca's aphasics' access to words is often more limited in free conversation than on testing with picture naming. They are aided on some words by being provided with the opening sounds. Because Broca's aphasics may have severe articulatory difficulties, their naming efforts may be hampered by articulatory struggle and distortion. They misname objects occasionally and use one-word circumlocutions when word retrieval fails. Their absolute level of naming impairment may range from complete failure to a mild reduction in speed of retrieval.

Wernicke's aphasics' anomia may be overshadowed by their fluent but noncoherent (paragrammatic) speech output. The point of their message may be so obscured by misused words that specific instances of word-finding failure may be hard to identify. Severe Wernicke's aphasics are also severely anomic on testing with pictures, and the anomia may abate considerably as the coherence of speech improves.

Example of a very severe Wernicke's aphasic with paragrammatism:
E: How are you today?
Pt: I feel very well. My hearing, writing been doing well; things that I couldn't hear from; in other words, I used to be able to work cigarettes I didn't know how. The pay I didn't know how. I can write, chesterfeela for over twenty years I can write it. Chesterfeel, I know all about it, I can write it.

Anomic aphasia is a pure disorder of word retrieval in the context of well-formed sentences, with little or no misuse of words. Anomic aphasic patients vary considerably as to how they cope with their frequent word-finding failures, which usually affect the nouns more than other parts of speech. Some patients barely miss a beat, providing circumlocutions as they speak, whereas others punctuate their discourse with expressions of frustration and impatience. Yet, in absolute levels, anomic aphasics may score with only mild impairments on standard naming tests. The prominence of their difficulty arises from their otherwise fluent normal speech and their efforts to use specific words.

Example of a severe anomic aphasic:
    E: Can you tell me about your illness?
    Pt: I had a . . . I had a . . . one or two three . . . There's one . . . I had a . . . a . . . I know the exact part of it.
    E: You're pointing to the operation on your chest?
    Pt: Yes, . . . I had a vaw . . . a lord . . . a w . . . It was replaced. It came back . . . I.
    E: A valve?
    Pt: Right . . . of the ower . . . the . . . the . . . There are three or four different things they could have in mind.
    E: Was it the aortic valve?
    Pt: Exactly.
    E: And it was after the operation?
    Pt: Right. About a day later, while I was under whatchamacall . . .
    E: Anesthesia?
    Pt: No. Under where they put you, just two or three people, an' you stay in there for a couple o' days.
    E: In the intensive care?
    Pt: Right. At that time I got the stroke.

Conduction aphasics have a fluent, grammatically organized output, punctuated by many episodes of going astray on the selection and ordering of the sounds of a word (phonological paraphasia), as well as going astray in the construction of a sentence. Both types of errors lead to self-corrective efforts. They characteristically make repeated unsuccessful attempts to straighten out the choice and order of sounds of a word. Phonological paraphasias are most likely to be confined to nouns, and may be compounded with severe word-retrieval difficulty. In other instances, it appears clear that the patient's efforts are based on an accurate unspoken model of the word's phonology, and that the problem is entirely one of production.

Example of a severe conduction aphasic:
    E: Tell me about what's going on in this picture.
    Pt: Oh . . . he's on top o' the ss . . . ss . . . swirl . . . it's a . . . ss . . . sss

. . . ss . . . sweel . . . sstool . . . stool. It's fallin' over. An' the girl, . . . the boy is . . . ss . . . 'ettin' his sister a . . . . He's. He'ss givin' her a ss . . . a . . . sss . . . sss . . . ssl . . . s . . . ss . . . sl cook . . . It's a soos . . . ss . . . ss . . . sss.

E: So he's giving his sister a cookie. What else is happening?

Pt: Well he he's fillin' out the ch ch ch . . . Oh, anyhow, his mother; is . . . she's bissy, but the water's fallin' over . . . The water is fallin'over the . . . the . . . er . . . the er . . . It's going flink . . . ss.

E: Yes, it's falling over the sink.

Pt: sull . . . sit . . . flink . . . er

E: Listen to me . . . sink.

Pt: Stink . . . sink . . . sink . . . sink . . .

E: OK, what is she doing there?

Pt: She's drawing the . . . she's drying the dishes.

The four syndrome types above were chosen as illustrations because the word-finding difficulty appears in a very different context of impairments in each of them. Although the absolute range of severity of the anomia covers the entire gamut in each of the syndromes, we can note certain characteristic differences. For example, conduction aphasics may be profoundly anomic, but a significant proportion of their efforts display partial retrieval of word phonology. Some anomic aphasics have access to a fairly large vocabulary, but their failures are absolute, with no partial phonological retrieval.

CLUES FROM ERROR PATTERNS IN APHASIC NAMING FAILURES

Naming failure for some patients appears as an inability to find the articulatory position for a word; for others it appears as the production of an off-target word; and for others as a near approximation of the desired word, which the patient tries unsuccessfully to correct. Some patients characteristically supply the intended word as soon as they are offered the first sound; others are not aided by such cues. Some patients are able to name colors but not objects, whereas some are able to name objects described to them, but not objects shown to them. These are but a few of the paradoxical observations that have constituted the pieces of the puzzle as to how name retrieval is realized by the brain. Another set of puzzle pieces is represented by the association of particular lesion sites with some of these error patterns. We will review systematically the clinical varieties of name-retrieval failures in aphasia, the lesion sites most regularly associated with them, and possible interpretations.

Initiating the Process: Visual Input

Naming to confrontation is the most common way of examining naming ability, and it usually corresponds to the patient's performance in conversation. It usually is assumed, correctly, that aphasic patients recognize objects that they see and appreciate their semantic properties. That is, visual agnosia is very rare in patients with aphasia. Thus, failures to name objects on sight usually begin with a malfunction further "downstream" within the language system proper. Yet, visually related naming disorders, though rare, are well documented and provide insight into the interface between the visual system and language. *Optic aphasia*, the impairment in name retrieval that is limited to visually presented objects, is always associated with an injury in the visual-association area of the left hemisphere, and often with degradation of the visual recognition process as well. In most cases of anomia, the naming deficit appears regardless of modality of the input, whether the object is seen, heard, or palpated. Modality-specific anomias will be discussed in detail in Chapter 4 by De Bleser, this volume.

Of course, the perceptual stimulus is only one way of initiating name retrieval. More commonly, it is initiated by a self-generated concept to be conveyed, or a question to be answered. This brings us squarely into the central question of the word-retrieval puzzle—how do we move from a concept to the retrieval of the phonological form of its name?

Failure to Activate Phonology

Simple failure to come up with a spoken response is the most common form of word-finding failure in aphasia. Depending on the patient's residual capacities, it may be accompanied by expressions of frustration or by circumlocutory comments. Most theorists agree that retrieval of word phonology begins after some level of semantic specification of the item to be named has been attained. Can the naming failure begin with some defect in the semantic specification of the concept? Some investigators claim that this is sometimes the case (Caramazza, Berndt, & Brownell, 1982; Whitehouse, Caramazza, & Zurif, 1978). Indeed, misnaming with semantically related words could be taken to imply semantic underspecification. But in most instances, the patient's own explanatory efforts and responses to probing questions are convincing evidence that his or her concept of the intended target is intact. This is particularly true in anomic aphasics, who may produce excellent pantomimes and informative circumlocutions, while being totally blocked on phonological retrieval. Behaviorally, we may describe this as a disconnection between semantics and phonology. Because both of these terms are constructs that do not have a known structural base, the notion of "disconnection" must be taken as describing a cognitive, and not an anatomical, disconnection.

Anomia with little or no impairment in other aspects of speech output has been reported in conjunction with three different lesion sites. One of these is the region of the temporo-parietal junction, particularly the angular gyrus; one is in the anterior periventricular region of the frontal lobe, and one is in the inferior temporal lobe. Severe anomia following temporal pole removal, without involvement of the classical language zone, also has been reported (Miceli, Giustolisi, & Caramazza, 1991). The differences in the clinical presentations of anomic symptoms from these different sites are subtle and not entirely predictable.

Anomia in conjunction with impaired comprehension of spoken words is found with lesions of the left superior temporal gyrus (in patients with Wernicke's aphasia), and with lesions of the temporo-occipital region (in patients with transcortical sensory aphasia). In the case of transcortical sensory aphasia it has been speculated that the lesion blocks the gateway into the language system for nonlinguistic semantic information. Chapter 3 by Tranel et al., this volume, provides insights into the neurological basis for name retrieval, as gleaned from the most current anatomical-functional studies.

TYPES OF PARAPHASIA

The variety of errors of partial retrieval and misretrieval have provided a rich basis for speculation concerning ways in which the naming system can malfunction. The term *paraphasia* refers to any type of unintended utterance, whether the error is at the level of phoneme choice or at the level of word choice; and whether recognizably related to an intended target or not.

Verbal Paraphasia

Errors of misnaming almost always involve substituting a semantically related word for the target, as in producing "green bean" for "asparagus," "seahorse" for "unicorn," or "Taj Mahal" for "sphinx." These are referred to as *semantic paraphasias*. However, verbal misnaming may occur with production of a word totally unrelated to the picture (e.g., "ball" in response to a picture of a visor; "chair" in response to a picture of a comb). Semantic paraphasias in response to pictures are made about as frequently by aphasic patients of all types. They often are immediately detected and rejected by the patient, but even when the patient is unaware of his or her error, one cannot assume that he or she is uncertain about the semantic features of the target object.

Perceptually Influenced Semantic Errors

Any type of associative link may lead to the activation of a semantically related paraphasia. An intriguing example is a response determined by some purely configurational relationship to the stimulus picture. For example, the stimulus "pinwheel" provoked the response by one patient as "a flower . . . a toy." The patient's reference to the object as a toy shows that he did not mistake the pinwheel for a flower, but was influenced by its configuration.

Phonemic (Literal) Paraphasia

Many responses reflect partial retrieval of word phonology but are partly erroneous. These sound-similar errors are referred to as *phonemic paraphasias.* Responses such as "ranno" for a rhinoceros, "spink" for sphynx, and "tums" for tongs, indicate that the patient has retrieved some features of the word's phonology, but not sufficiently to completely guide the production of the target name. Sometimes a response may appear to be an awkward production of a correctly targeted effort. For example, is the response "breestace," followed by "breegstayst" an error of sound choice or an awkward misproduction of the target briefcase? In the latter case, it would better be classified as an articulatory error, rather than as a phonemic paraphasia.

Among the most interesting phonemic paraphasias are those that are subjected to repeated self-corrections. Responses typical of conduction aphasics suggest that the patient may have an accurate phonological template in subvocal form, but cannot organize his or her output to realize it. If this model were accurate it would be congruent with Wernicke's classical anatomical disconnection explanation for conduction aphasia, as endorsed by Geschwind (1965). According to this view, the phonological memory for words, which is retrieved in Wernicke's area (the superior temporal gyrus) is normally conveyed to Broca's area for articulatory realization, through a white matter pathway (the *arcuate fasciculus*). In conduction aphasia the anatomical model postulates an interruption of this communicating pathway, resulting in imperfect or failed transmission of phonology to the articulatory system.

This view has not stood up to close clinical, experimental, or anatomical scrutiny in the last 20 years. Conduction aphasics provide evidence of phonological knowledge in only a minority of words that they fail to articulate (Goodglass, Kaplan, Weintraub, & Ackerman, 1976). The clinical disorder characterized by a predominance of phonemic paraphasias with good auditory comprehension has been found to be associated with lesions at several points in the perisylvian cortex, making it unlikely that this is a disconnection phenomenon (Damasio & Damasio, 1989).

Neologistic Paraphasia

Although the patient's response may carry undeniable evidence of guidance by the phonology of the intended target, it may be so contaminated with extraneous sounds that the result must be labeled *neologistic*. Phonological segments carried over from the target need not even respect the position they occupied in the intended utterance. For example, the picture of a harmonica led to the production "con . . . caca, commat . . . mon . . . no!" (Goodglass, 1993). In many cases, neologistic utterances contain no trace of the target-word sounds. Neologistic attempts are common in patients with severe conduction aphasia or Wernicke's aphasia. Neologisms occasionally are produced by Broca's aphasics as well. (Anomic aphasics virtually never produce either phonological or neologistic paraphasias.)

Phonosemantic Blends

The associative processes that may lead astray the realization of an attempt to name may show multiple influences. Some errors appear to be determined by both semantic and phonological influences. For example, "brush," in response to a picture of a broom, or "eskloot" in response to a picture of an "igloo" (Goodglass et al., 1997).

Failures of Articulation

The classical anatomical model suggests that many patients should fail to emit the names of objects purely on the basis that they cannot generate or implement the articulatory plan, although all the "upstream" stages of naming are intact (viz., picture recognition and semantic interpretation, retrieval of word phonology in Wernicke's area, and transmission of information to the articulatory system). These should be patients with a disorder primarily in articulatory planning: Broca's aphasics in the classic terminology. Indeed, there is no lack of evidence of patients who struggle with impaired articulation to produce a barely intelligible version of an object name, but one that must be credited as correct, making allowance for the patient's production impairment. We saw above the example of "breegstast" for briefcase. Other examples were "spink" for sphynx and "montigah" for harmonica.

It is more difficult, however, to be convinced that a patient's failure lies only in the articulatory planning when he or she cannot initiate a response at all or begins with a totally extraneous sound. Goodglass et al. (1976) found that Broca's aphasics did not differ from chance in providing evidence of tacit phonological knowledge of a word that they could not name. One may ask, "What about patients who can write words that they cannot say aloud?" Friederici, Schönle, and Goodglass (1981) studied the written responses of Broca's aphasics who could write more words than they could name. Their results favored a "dual coding" view of written and oral naming. That is, the errors made by these patients were semantic substitutions, rather than structural spelling errors. The patients appeared to be writing from a graphic system that did not depend on having access to the phonology of the target. We cannot assume that being able to write an object name after failing to retrieve it in speech means that the patient "knows the word" and just cannot pronounce it.

Perseveration

The inappropriate repetition of a response to an earlier stimulus is a common phenomenon in many domains of performance by brain-damaged patients. Perseverative errors in naming are particularly common in aphasic patients. Albert and Sandson (1986) proposed distinguishing between repetition of an immediately preceding response (stuck-in-set perseveration) and repetition of an earlier, but not immediately preceding one (recurrent perseveration). But the persisting influence of earlier behavior on a new response need not be a complete reproduction of the earlier response. Inappropriately carrying over part of the phonology of a previous word in a new response is one of the common forms of perseverative naming behavior, and is even more frequent than repeating an earlier response because of a semantic link (Helm-Estabrooks, Emery, & Albert, 1987). One of our patients' records show the following:

Stimulus picture	Response comment
broom → "brush"	Phonosemantic blend
hanger → "brush"	Stuck-in set perseveration
saw → "not scissors, but . . ."	Recurrent perseveration. (Scissors was correctly named earlier in test.)

Perseverative reactions may inhibit the correct response by being stronger competitors because of their recency than a correct target that is difficult to retrieve. Helm-Estabrooks et al. (1987) found that treatment to inhibit perseveration in naming actually increases the performance of patients upon retesting. Helm-Estabrooks reviews the treatment of aphasic perseveration, as well as other therapeutic techniques that have been developed for word-retrieval problems in Chapter 8, this volume.

Misperceptions

Every examiner has had patients misname pictures because of a perceptual misinterpretation. The patient offers a name that is correct for the object he or she thinks the picture represents. Such misperceptions are common in elderly demented subjects, less so in normal elderly, and only occasional in aphasic patients. Some examples from our aphasic population are "lollipop" for a baby's rattle and "a knot" for a pretzel. We do not consider them naming errors.

Circumlocutions

Many patients compensate for a word-retrieval failure by telling something about the object, in lieu of naming it. Such circumlocutions are much more commonly given by anomic aphasics than other types of patients, largely because they have the fluent access to alternative expressions that other patients lack. There is, however, a type of one-word response that is in fact a circumlocution, though often incorrectly scored as verbal misnaming. When the patient responds to the picture of an object by giving a verb or an adjective, it is clear that he does not think he is naming the object. The responses "smoking" for a cigarette, "drip drip" for a nozzle, and "electric" for a plug, may be given by patients of any type, including Broca's aphasics. It is possible that patients may attempt to use such circumlocutions as self-prompting devices, much as a normal speaker in a tip-of-the-tongue state may verbalize related, but incorrect words, in the hope of being led to the desired response.

EXPLANATORY MODELS

Coherent efforts to account for the phenomena of aphasia in terms of cause and effect probably began with Wernicke (1874). Wernicke proposed that the temporal lobe contained a store of auditory word forms, and it was connected to a conceptual store, from which the sounds of words drew their meaning. Disconnection of the auditory word center from the conceptual store would prevent the selection of a word when a concept arose or was activated by an external stimulus. The correct articulatory production of the word by the motor speech center depended on its receiving the information about the auditory image from the temporal lobe. Interruption of the connecting pathway would result in an output of incorrect sounds. In Geschwind's (1969) adaptation of Wernicke's model, visual form is perceived in the occipital lobe, and is transferred to the region of the left angular gyrus. The angular gyrus communicates with Wernicke's area in the temporal lobe, where its phonological form is generated. This information is conveyed via the arcuate fasciculus to Broca's motor speech area, where a motor plan is generated, to be executed by the motor cortex. This is the prototype of a serial stage model, in that information is passed along from center to center, initiating a different set of operations at each center.

The anatomically based serial stage model had considerable support from clinical observation. For example, the existence of optic aphasia, as described above, confirms that object naming requires intact visual input to the language system. Geschwind's assignment of a major role in name retrieval to the left angular gyrus also had support from both the anatomic and clinical sides. Anatomically, the angular gyrus is a convergence zone for connections from visual, auditory, and tactile areas of the brain. Geschwind argued that the development of this convergence zone in humans is what makes it possible for us to acquire names for objects. Clinically, lesions in the angular gyrus are a common source of anomic aphasia, with little impact on other aspects of oral language.

Geschwind was not alone among modern students of aphasia to make the temporal lobe a critical zone for the activation of the auditory memories of words of the language. Luria (1970) also held that lesions of the temporal lobe disrupt the auditory representation of words and result in severe word-retrieval difficulty. Indeed, patients with Wernicke's aphasia following injury to the superior temporal gyrus are usually severely anomic.

The postulation that it is the arcuate fasciculus that carries phonological information from the temporal lobe to the motor articulatory planning center in Broca's area derives primarily from the assumption, in the anatomical model, that there must be such a connection. Given that assumption, the identification of the arcuate fasciculus as the pathway that served that function was dictated by the anatomical fact that the arcuate fasciculus occupied a location that would permit it to carry out that postulated function. Many cases of conduction aphasia were due to lesions of the supramarginal gyrus so placed as to damage the arcuate fasciculus. Given the bias induced by the assumptions of the anatomical model, this constellation of observations appeared to be an unshakable confirmation of those assumptions.

The final stage in the anatomical model concerns the role of Broca's area as the center that merely implements the articulation of the linguistic information that comes its way. The association of Broca's area lesions with disorders primarily affecting the articulation of speech was well known, as was the fact that many patients with a lesion confined to this zone had remarkably intact comprehension of speech. The notion that their word retrieval was damaged only at the level of motor implementation was shared even by Luria (1970). Luria cited the superior ability of Broca's aphasics to complete the opening sounds of an object name after they had failed to retrieve the word without assistance. He felt that this was evidence of a well-preserved auditory word image, due to the intactness of the temporal lobe.

The transparent logic of a naming process that moves in steps from visual input to motor articulatory output also has inspired naming models in cognitive psychology—models that do not necessarily make any reference to the brain. For example, Levelt (1989) proposed a well-developed serial stage model that has the following stages: The first stage represents the recognition of a pictorial stimulus. The next stage is the semantic specification of the word to be used, including its unique configuration of properties and its syntactic demands. The product of this stage is referred to by Levelt as the "lemma." The lemma, per se, does not include the phonological form of the word, but it is a pointer to the location of the phonological information in an internal lexicon. Following the suggestion of Dell (1986), the actual assembly of phonology proceeds through a hierarchy of nested informational units, with the word node leading to a succession of syllables, and each syllable unfolding from "left to right," to complete a phonetic plan. The phonetic plan is then passed on to the articulatory realization system for implementation. The parallel between a model based in anatomy and one founded on purely cognitive considerations can be striking, as seen in Figure 2.

Serial stage models meet the ideals of a good theory: They organize a

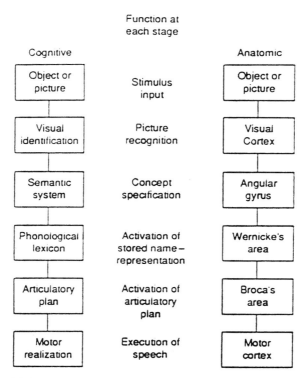

Figure 2   Parallels between cognitive and anatomic serial stage models of object naming. (From Goodglass, 1993, Fig. 5.1, p. 95).

great deal of data into a plausible framework. They are sufficiently brittle to be falsifiable. They lead to testable experimental hypotheses. To this degree, it can be said that they serve as a valuable heuristic in studies of object naming (Wingfield & Wayland, 1988). In the preceding discussion we pointed out the extent to which the anatomically based serial stage theory organizes existing clinical and anatomical observations.

Beyond the Borders of Serial Stage Models

Serial stage models owe their wide acceptance both to their logical appeal and to the fact that they organize a great many observations into a coherent account. Yet the observations that are so organized are a restricted set. Some of these observations are taken as correct because they fit the logic of the model, but this occurs at the expense of ignoring equally common observations that do not fit the model. In the next sections, we will review clinical and experimental data that are not accounted for, or that may contradict serial models.

The Microstructure of Access to Name Phonology

An assumption of serial stage models, both anatomic and cognitive, is that the state of the partially completed name activation can be specified at each point. That is, a lesion in a given stage should find particular aspects of name retrieval accomplished when they are tested for with appropriate probes. We have referred earlier to the use of word-onset phonemic priming as a probe to see whether the patient has retrieved enough of the word phonology to be able to complete an object name that he or she has failed to name on picture confrontation. By this we mean that when the patient has failed to name an object, the examiner provides the beginning sounds of the target word. Most observers (Goodglass & Stuss, 1979; Luria, 1970) concur in finding that Wernicke's aphasics, with their temporal lobe lesions, are significantly less responsive to these primes than are Broca's aphasics. This observation would be consistent with the belief that phonological activation of words takes place in Wernicke's area and is most vulnerable to a lesion at that site. However, Goodglass and Stuss reported that anomic aphasics with lesions in the angular gyrus area also respond well to word-onset phonemic primes. In the serial stage model described by Geschwind, however, the angular gyrus is a way station that precedes the temporal lobe in the processing sequence. Anomia produced by an angular gyrus lesion should prevent the activation of word phonology in the temporal lobe and hence should result in poor response to phonemic priming. A study by Wingfield, Goodglass, and Smith (1990) casts some doubt on whether phonemic priming is a reliable clue to the state of tacit phonological activation resulting from a patient's effort to retrieve a picture name. They questioned whether the completion of the opening segment of a word might not simply reflect the activation of familiar phonological string, independently of whether it was the target of a word-retrieval effort. The empirical background for this question came from the work of Grosjean (1980) and Marslen-Wilson (1984), showing that with normal subjects, words heard within sentence contexts can be recognized, on average, within as little as the first 200 msec of their onset, or when less than half of their full acoustic signal has been heard. Spoken words heard in isolation from any context require, on average, only 130 msec more. As surprising as this may seem, it derives from the fact that the average number of words in English that share the same phonemic onsets reduces dramatically as more and more of a word onset is heard (Tyler, 1984; Wayland, Wingfield, & Goodglass, 1989). When Wingfield et al. (1990) applied this method to aphasic subjects and normal controls, they found that the aphasics needed little more phonological word-onset information (358 msec) than normals (297 msec) in order to complete the object name. Although this difference was significant, it is clearly a small one in absolute terms, amounting to just 71 msec—or an average of only 1.2 additional increments of 50 msec—necessary for word recognition by the aphasic subjects as compared with the normal controls. That is, even patients with profound word-retrieval difficulties in picture-naming tasks were able to give correct completions to gated stimuli without prior knowledge of the target.

Figure 3 shows the average gate size (i.e., word-onset information in msec) needed by aphasics to give the correct full name after a naming failure, and by these same subjects (and the group of normal controls) when only word onsets were presented and no picture was present. The difference between the amount of word onset necessary for correct name production for the aphasics with (282 msec) and without the presence of the object picture (358 msec) is quite small in absolute terms. The mean difference in gate size is only 76 msec, or an average of just over 1.5 50-msec increments in gate size for a correct response. Word-onset phonology alone can thus be sufficient in many cases to evoke the target name through a strictly auditory route (cf. Wayland et al., 1989; Wingfield et al., 1990).

The study of Wingfield et al. (1990) indicates that the phonemic priming technique may not be an appropriate probe of the activation produced by the effort to name a picture. Other techniques, however, also have failed to support the assumption that the level of partially attained phonological knowledge can be specified at each lesion site. As indicated earlier, the assumption that, in Broca's aphasia, phonological activation is complete up to the point of articulatory realization, is contradicted by the study of

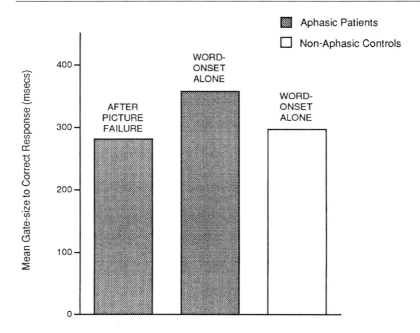

Figure 3 Mean amount of word-onset information (gate-size in milliseconds) necessary for recognition of spoken object names by aphasic patients after failing to name the object picture or when just hearing the spoken names alone with no picture present. Data are also shown for normal subjects who heard the same names spoken with no pictures present. (Data from Wingfield, Goodglass, & Smith, 1990.)

Goodglass et al. (1976), in which Broca's aphasics were no better than chance in selecting the first letter of words that they failed to retrieve, or in indicating the number of syllables in it. The anatomic model holds that conduction aphasics also have a lesion to the system after the point of phonological activation and that their speech-output difficulties are due to a disconnection of the temporal lobe from motor articulatory centers. The expectation of tacit phonological knowledge on the part of conduction aphasics was partly supported by the findings of Goodglass et al. (1976). Conduction aphasics were the only group who performed at better than chance levels in demonstrating awareness of initial sounds and of syllable count; however, they did so successfully on only one-third of the pictures that they failed to name. Thus the data from systematic observation contradict a basic axiom of the anatomic stage model.

Immediate versus Delayed Lexical Retrieval

Current serial stage models of word retrieval deal only with normal, relatively rapid access to the phonological form. The process of retrieving a word through prolonged search or after giving up active retrieval efforts has been the focus of investigations of the "tip-of-the-tongue" (TOT) state. The TOT state refers to cases of word-retrieval failure in which the person is confident that he or she knows the word, but where the word remains frustratingly out of reach. Brown and McNeill (1966) are credited with the first systematic study of the effect, in which they used subjects' partial knowledge to make inferences about their retrieval strategies. A good review of TOT studies with normal subjects can be found in Brown (1991).

In studying the relationship between word frequency and response latency, Goodglass, Theurkauf, and Wingfield (1984) found that there was a break in the correlation between these two factors at about 1500 msec. That is, the well-known inverse relationship between naming latency and word frequency for both normal subjects (Oldfield & Wingfield, 1965; Wingfield, 1967, 1968) and aphasic patients (Newcombe, Oldfield, & Wingfield, 1965; Newcombe, Oldfield, Ratcliff, & Wingfield, 1971) dropped to a random relationship for object pictures that took longer than 1500 msec to name. Goodglass et al. (1984) proposed that this was the point at which subjects became aware of their failure to supply a word through normal rapid processing and might attempt conscious associative strategies to retrieve the elusive target word.

Two opposing positions have developed as to how prolonged efforts at word retrieval in the TOT condition achieve eventual success. Some investigators report that a deliberate strategy of thinking of words that are associated to the target often leads to successful retrieval. The alternative theory holds that successful retrieval occurs as the result of a continuing search process that is outside of awareness and outside of conscious control. Kohn et al. (1987) offered evidence favoring the view that subjects' conscious efforts had no influence on eventual word retrieval in a TOT state. In a related study, Goodglass, Wingfield, and Wayland (1989) found that continuous word-retrieval efforts during a prolonged (30-sec) TOT state did not bring a word any closer to retrieval threshold than such efforts during a short (10-sec) one. An advantage favoring a 30-sec search time over the shorter interval would have favored a progressive search model. The finding of no difference fails to support the progressive search hypothesis (although it cannot, of course, directly prove the alternative one). Although the observations summarized here may be reconcilable with a serial stage model, there has been no effort to date to make such an attempt.

Category-Specific Dissociations

Serial stage models (whether anatomically or purely cognitively based) treat all name retrieval as following the same path. They start with visual recognition, proceed to concept specification (or lemma choice), then to phonological retrieval, and finally to articulatory planning and implementation. The discovery of selective impairments or selective sparing of certain categories of words posed a challenge to theorists, which seems to require stepping outside of the framework of a serial naming model. Goodglass, Klein, Carey, and Jones (1966) first reported selective sparing for the retrieval of letter names, along with other category-specific deficits in aphasia. Goodglass, Wingfield, Hyde, and Theurkauf (1986) added body parts and colors to the list of categories that were often relatively spared in fluent aphasics with anomia.

The categories affected by selective dissociations in aphasia were almost all members of a tightly associated, limited set. However, Yamadori and Albert (1973) described a patient who could not comprehend the names of "indoor objects," such as furniture and parts of a room. Beginning with the paper by Warrington (1975), Warrington and her associates described a series of cases of lexical dissociation affecting auditory comprehension and/or production where the categories affected demanded description in broader semantic terms, such as abstract versus concrete, or "naturally occurring" versus "man-made" (Warrington & McCarthy, 1983, 1987; Warrington & Shallice, 1984.

Two cases of selective anomia that were reported by Rapcsak, Comer, and Rubens (1993) and Rapcsak, Kaszniak, and Rubens (1989) are particularly instructive because of the narrow category affected (words denoting facial emotional expressions), the limitation of the deficit to response to visual confrontation, and the specific site of the causative lesion. The site in these cases was the inferotemporal region of the *right* hemisphere. These cases are the only ones, to our knowledge, that document a specific anatomical site associated with a particular semantic content, and suggest a mechanism that may be unique for this particular dissociation. Rapszak et al.'s (1993) explanation for the mechanism relies on observations by Ojemann, Ojemann, and Lettich (1992) that microelectrode recordings from the right temporal lobe revealed cortical units that became more active during the naming of facial emotions. The same authors found face-specific cells in the inferotemporal cortex of the nondominant hemisphere in humans, complementing multiple observations of regions of face-specific and facial-expression-specific neurons in the inferotemporal cortex of monkeys.

The observations on category-specific naming dissociations should be taken in the context of the more general observations of anomia in aphasia. The categories that are subject to selective dissociations are a limited set. Those that typically have been selectively impaired in patients with postencephalitic lesions of the inferior temporal lobe have been animals (Hart & Gordon, 1992), fruits, and vegetables (Hart, Berndt, & Caramazza, 1985; Sartori & Job, 1988), and they have been contrasted with manipulable objects, such as tools and utensils, which are usually spared in these patients. Warrington and McCarthy (1987) studied an aphasic patient with a pattern of selective impairments that was the reverse of the one commonly observed with inferotemporal lesions (i.e., they showed superior performance with animate objects and severe impairment for inanimate ones). By examining a great many categories of both picturable and nonpicturable nouns, Warrington and McCarthy found that many types of nouns did not conform to the animate—human-made dichotomy. For example, body parts, which could be presumed to be animate, clustered with the well-preserved categories of human-made objects. Processed foods (bread, cheese, etc.) generally have been reported as suffering the fate of animate objects in postencephalitic patients. In Chapter 3, this volume, on the neurology of naming, Tranel, Damasio, and Damasio pursue the issue of anatomic localization specific to a number of the category-specific anomias. Of course, a full consideration of the anatomy underlying impairments of word retrieval will include both observations from patients with word-retrieval problems that are of a general type, and those related to category-specific problems.

Linguistically Based Categories

There are two classes of dissociation for name retrieval that must be defined using linguistic, rather than semantic, categories. These are impairments in the use of proper nouns and differential impairment of noun and verb production. The latter dissociation is of special interest, because agrammatic patients usually access nouns more easily than verbs, whereas the opposite is true for anomic aphasics. Although problems in recalling proper nouns are a common complaint of normal elderly people (e.g., Burke et al., 1991), selective impairments of this type are quite rare in aphasia. Indeed, the patient described by Semenza and Zettin (1989) with a marked proper name anomia, was not aphasic in other respects. Chapter 5 by Semenza covers both the empirical data and the theoretical considerations that may underlie this phenomenon.

Difference in competence for using nouns, in comparison with the use of verbs, arises repeatedly in aphasic patients of various types. These differences may appear in spelling, naming, and word retrieval. Caramazza and Hillis (1991), for example, recently contrasted two patients, one of whom could produce verbs orally, but not in writing, whereas the other could write verbs but not say them aloud. Caramazza and Hillis (1991) compared their first patient in her ability to write the same lexical term (e.g., "crack") in its use in a sentence as a noun, as opposed to its use as a verb. In spite of the identical form of the word in its two uses, the patient could write it only in its noun function. Caramazza and Hillis (1991) suggested that nouns and verbs are organized in different lexicons that may be separately impaired.

It has long been noted that nouns are sparse in the speech of anomic aphasics, but that verbs are relatively accessible to them. The reverse is the case for agrammatic patients. Kohn (1989) and Miceli, Silveri, Villa, and Caramazza (1984) showed that the verb-retrieval deficit that is characteristic in the spontaneous speech of many patients with agrammatism was also apparent in tests of single-word retrieval using pictures of objects (nouns) versus actions (verbs). Goodglass (1993) proposed that the differences in access to nouns and verbs may lie in the contrast between the act of labeling and the act of predicating. In some forms of aphasia—specifically agrammatism (Luria, 1970)—the underlying deficit of the patient may be in the act of predicating. Verbs are intrinsically predicative terms, hence the same lexical term (e.g., "hammer") may be there for the agrammatic speaker to use as a label, but not as a verb. Serial stage modeling has been uninformative in dealing with any of the dissociative phenomena described in these preceding sections.

Alternatives to Serial Stage Models

With the increasing sophistication of computer modeling, an increasing number of theorists have turned to various forms of parallel, interactive models to understand language and other cognitive processes. The essential feature of these approaches is that seemingly "rule-governed" behavior may be an emergent consequence of the interaction of a large number of elementary associative connections. Unlike the primitive elements of serial stage models, those in connectionist networks do not have names and clearly diagrammable effects on other named systems. Hence they lack the transparent logic of linear stage models. They must be implemented in a computer to see if they produce an output like that of the natural behavior that they are meant to model.

ORGANIZATION OF THIS VOLUME

The remainder of this volume is organized into three sections:
Part II: "Anatomical and Theoretical Considerations in Anomia" con-

sists of two chapters. Gordon approaches the theory of naming from the cognitive viewpoint and points to the areas in which cognitive models can be integrated with anatomical findings, as well as those in which the linkage is still to be explained. Tranel, Damasio, and Damasio offer an integration of lesion and functional imaging data with an anatomical view of the name retrieval system.

Part III: "Dissociations and Other Naming Phenomena" consists of two chapters. De Bleser, in her chapter on modality-specific anomia, discusses the evidence for word-retrieval problems that are confined to a particular sensory input or motor output channel, and their possible underlying mechanisms. Semenza focuses on the isolated loss of retrieval of proper names, treating it primarily in terms of its cognitive relationships.

Part IV: "Life-Span Perspectives on Anomia: Clinical and Therapeutic Considerations" consists of three chapters. Menyuk deals with naming disorders in the early childhood phase, both in children with early brain injury and in children with purely developmental abnormalities. Nicholas, Barth, Obler, Au, and Albert present an integration of behavioral observations of changes in naming functions in normal and dementing older individuals. In the final chapter, Helm-Estabrooks presents the rationale and methodology of various therapeutic approaches to naming disorders, along with an assessment of results.

REFERENCES

Albert, M. L., & Sandson, J. (1986). Perseveration in aphasia. *Cortex, 22,* 103–115.

Alexander, M. P., Fischette, M. R., & Fischer, R. (1989). Crossed aphasia can be mirror image or anomalous. *Brain, 112,* 953–973.

Brown, A. (1991). A review of the tip-of-the tongue experience. *Psychological Bulletin, 109,* 204–223.

Brown, R., & McNeill, D. (1966). The "tip of the tongue" phenomenon. *Journal of Verbal Learning and Verbal Behavior, 5,* 325–337.

Burke, D. M., MacKay, D. G., Worthley, J. S., & Wade, E. (1991). On the tip of the tongue: What causes word finding failures in young and older adults? *Journal of Memory and Language, 30,* 542–579.

Caramazza, A., Berndt, R. S., & Brownell, H. (1982). The semantic deficit hypothesis: Perceptual parsing and object classification by aphasic patients. *Brain and Language, 15,* 161–189.

Caramazza, A., & Hillis, A. E. (1991). Lexical organization of nouns and verbs in the brain. *Nature, 349,* 788–790.

Chertkow, H., & Bub, D. (1990). Semantic memory loss in dementia of the Alzheimer's type. *Brain, 113,* 397–417.

Damasio, H., & Damasio, A. R. (1989). *Neuropsychological disorders: Neuroimaging procedures and problems.* New York: Oxford University Press.

Dell, G. (1986). A spreading activation theory of retrieval in sentence production. *Psychological Review, 93,* 283–321.

Friederici, A. D., Schönle, P. W., & Goodglass, H. (1981). Mechanisms underlying writing and speech in aphasia. *Brain and Language, 13,* 212–223.

Geschwind, N. (1965). Disconnexion syndromes in animals and man. *Brain, 88,* 237–294, 585–644.

Geschwind, N. (1969). Problems in the anatomical understanding of the aphasias. In A. L. Benton (Ed.), *Contributions to Clinical Neuropsychology* (pp. 107–128). Chicago: Aldine.

Goodglass, H. (1993). *Understanding Aphasia.* San Diego: Academic Press.

Goodglass, H., Kaplan, E., Weintraub, S., & Ackerman, N. (1976). The "tip-of-the-tongue" phenomenon in aphasia. *Cortex, 12,* 145–153.

Goodglass, H., Klein, B., Carey, P., & Jones, K. J. (1966). Specific semantic word categories in aphasia. *Cortex, 2,* 74–89.

Goodglass, H., & Stuss, D. T. (1979). Naming to confrontation versus oral description in three subgroups of aphasics. *Cortex, 15,* 199–211.

Goodglass, H., Theurkauf, J. C., & Wingfield, A. (1984). Naming latencies as evidence for two modes of lexical retrieval. *Applied Psycholinguistics, 5,* 135–146.

Goodglass, H., Wingfield, A., Hyde, M. R., Gleason, J. B., Bowles, N. L., & Gallagher, R. E. (1997). The importance of word-initial phonology: Error patterns in prolonged naming efforts by aphasic patients. *Journal of International Neuropsychological Society, 3,* 1–11.

Goodglass, H., Wingfield, A., Hyde, M. R., & Theurkauf, J. C. (1986). Category-specific dissociations in naming and recognition by aphasic patients. *Cortex, 22,* 87–102.

Goodglass, H., Wingfield, A., & Wayland, S. (1989). The nature of prolonged word-finding. *Brain and Language, 36,* 411–419.

Grosjean, F. (1980). Spoken word recognition processes and the gating paradigm. *Perception and Psychophysics, 82,* 267–283.

Hart, J., Berndt, R. S., & Caramazza, A. (1985). Category-specific naming deficit following cerebral infarction. *Nature, 316,* 439–440.

Hart, J., & Gordon, B. (1992). Neural subsystems for object knowledge. *Nature, 359,* 60–64.

Helm-Estabrooks, N., Emery, P., & Albert, M. L. (1987). Treatment of Aphasic Perseveration (TAP) program. *Archives of Neurology, 44,* 1253–1255.

Kohn, S. E. (1989). Verb finding in aphasia. *Cortex, 25,* 57–69.

Kohn, S. E., Menn, L., Wingfield, A., Goodglass, H., Gleason, J. B., & Hyde, M. R. (1987). Lexical retrieval: The tip-of-the-tongue phenomenon. *Applied Psycholinguistics, 8,* 245–266.

Levelt, W. J. M. (1989). *Speaking.* Cambridge, MA: MIT Press.

Luria, A. R. (1970). *Traumatic aphasia.* The Hague: Mouton.

Marslen-Wilson, W. D. (1984). Function and process in spoken word recognition. In H. Bouma & D. G. Bouwhuis (Eds.), *Attention and Performance X* (pp. 125–150). Hillsdale, NJ: Erlbaum.

Miceli, G., Giustolisi, L., & Caramazza, A. (1991). The interaction of lexical and non-lexical processing mechanisms: Evidence from anomia. *Cortex, 27,* 57–80.

Miceli, G., Silveri, M. C., Villa, G., & Caramazza, A. (1984). On the basis of agrammatics' difficulty in producing main verbs. *Cortex, 20,* 207–220.

Newcombe, F., Oldfield, R. C., & Wingfield, A. (1965). Object naming in dysphasic patients. *Nature, 207,* 1217–1218.

Newcombe, F., Oldfield, R. C., Ratcliff, G. G., & Wingfield, A. (1971). The recognition and naming of object-drawings by men with focal brain wounds. *Journal of Neurology, Neurosurgery, and Psychiatry, 34,* 329–340.

Ojemann, J. G., Ojemann, G. A., & Lettich, E. (1992). Neuronal activity related to faces and matching in human right nondominant temporal cortex. *Brain, 115,* 1–13.

Oldfield, R. C., & Wingfield, A. (1965). Response latencies in naming objects. *Quarterly Journal of Experimental Psychology, 17,* 273–281.

Rapscak, S. Z., Comer, J. F., & Rubens, A. B. (1993). Anomia for facial expressions: Neuropsychological mechanisms and anatomical correlates. *Brain and Language, 45,* 233–252.

Rapcsak, S. Z., Kaszniak, A. W., & Rubens, A. B. (1989). Anomia for facial expressions: Evidence for a category-specific visual-verbal disconnection syndrome. *Neuropsychologia, 27,* 1031–1041.

Sartori, G., & Job, R. (1988). The oyster with four legs: A neuropsychological study of the interaction of visual and semantic information. *Cognitive Neuropsychology, 5,* 105–132.

Semenza, C., & Zettin, M. (1989). Evidence from aphasia for the role of proper names as pure referring expressions. *Nature, 342,* 678–679.

Tyler, L. (1984). The structure of the initial cohort: Evidence from gating. *Perception and Psychophysics, 36,* 417–427.

Warrington, E. K. (1975). The selective impairment of semantic memory. *Quarterly Journal of Experimental Psychology, 27,* 635–657.

Warrington, E. K., & McCarthy, R. (1983). Category-specific access dysphasia. *Brain, 106,* 859–878.

Warrington, E. K., & McCarthy, R. (1987). Categories of knowledge: Further fractionation and an attempted integration. *Brain, 110,* 1273–1296.

Warrington, E. K., & Shallice, T. (1984). Category-specific semantic impairment. *Brain, 107,* 829–854.

Wayland, S. C., Wingfield, A., & Goodglass, H. (1989). Recognition of isolated words: The dynamics of cohort reduction. *Applied Psycholinguistics, 10,* 475–487.

Wernicke, C. (1874). *Der aphasische Symptomenkomplex.* Breslau: Cohn and Weigert.

Whitehouse, P., Caramazza, A., & Zurif, E. B. (1978). Naming in aphasia: Interacting effects of form and function. *Brain and Language, 6,* 63–74.

Wingfield, A. (1967). Perceptual and response hierarchies in object identification. *Acta Psychologica, 26,* 216–226.

Wingfield, A. (1968). Effects of frequency on identification and naming of objects. *American Journal of Psychology, 81,* 226–234.

Wingfield, A., Goodglass, H., & Smith, K. (1990). Effects of word-onset cueing on picture naming in aphasia: A reconsideration. *Brain and Language, 39,* 373–390.

Wingfield, A., & Wayland, S. C. (1988). Object naming in aphasia: Word-initial phonology and response activation. *Aphasiology, 2,* 423–425.

Yamadori, A., & Albert, M. L. (1973). Word category aphasia. *Cortex, 9,* 112–125.

Anatomical and Theoretical Considerations in Anomia

Both anatomically based and psychologically based accounts of the naming process and its impairments have come a long way since the explanations of the proposals of the classical schools of the late 19th and early 20th centuries. Anatomically oriented models, based on those of Wernicke (1874) and Lichtheim (1884) centered around the existence of a storehouse of auditory word images in the temporal lobe. Cognitively oriented proposals, such as those of Jackson (1866), Head (1926), and Goldstein (1948) were concerned with an impairment in the use of symbols for communication, expressed in terms of a loss of "propositionizing" or of "symbolic formulation and expression" or of "abstract behavior."

Anatomo-clinical approaches were handicapped by the appearance of a number of different lesion sites that might be candidates for the zone most vital to normal naming. Investigators lacked the technology for studying the functional anatomy of the normal naming process, so that these observations could be related to the effects of brain lesions.

The psychological accounts were in essence intuitive in nature and carried the notion of symbolization no further than showing how a philosophical construct had a counterpart in brain function. Psychological approaches to naming now have the benefit of cognitive techniques for examining processes as they unfold in real time. Furthermore, studies in depth of critically important cases involving selective dissociations of naming functions have produced approaches to performance analysis that were unknown a generation ago.

In the first of the two chapters in this section, Barry Gordon illustrates the power of modern cognitive theories as a framework for integrating anatomical and functional components of the naming process. In the following chapter, Daniel Tranel, Hanna Damasio, and Antonio Damasio bring to bear the results of new lesion studies and state-of-the-art functional imaging studies on one of the most fundamental issues in anomia: impairments in basic object naming (nonunique concrete entities), impairments in retrieval of proper names (unique concrete entities), and impairments in retrieval of action names.

REFERENCES

Goldstein, K. (1948). *Language and language disturbances.* New York: Grune and Stratton.

Head, H. (1926). *Aphasia and kindred disorders of speech.* New York: Macmillan.

Jackson, J. H. (1866). Notes on the physiology and pathology of language. *Medical Times and Gazette, 1,* 659.

Lichtheim, O. (1884). On aphasia. *Brain, 7,* 443–484.

Wernicke, C. (1874). *Der aphasische Symtomenkomplex.* Breslau: Cohn und Weigert.

Chapter 2

Models of Naming

Barry Gordon

INTRODUCTION

Visual confrontation naming—the focus of this chapter—is more than just a standard clinical tool and research task. It has many virtues that help make it a critical instrument for understanding many aspects of human cognition and their neural basis.

The ability to refer to objects by names may be at the root of human language development, in phylogeny as well as in ontogeny (Terrace, 1985). Yet naming is a relatively straightforward cognitive operation, whose outlines are well understood (despite the controversies that will be pointed out in this chapter!). Naming uses only a limited number of cognitive processing stages, and the nature of these stages is fairly well known. Processing through these stages is certainly sequential, in some sense. It is almost all feedforwards; it is not necessary to postulate internal feedback loops between stages. Experimentally, both the inputs and the outputs of the naming process are well defined. Complicating issues such as response buffering, processing strategies, memory and learning, syntax, and high-level concept formation can generally be ignored. Naming is also related to other basic abilities—word production, face recognition, and word reading—for which there is a wealth of data and productive theorizing.

Naming is known to be the product of relatively specific brain regions, and relatively small ones (perhaps on the order of several cm² in each individual), so the neural mechanisms involved must be comparatively circumscribed. Moreover, it is now possible to make detailed computational models of a wide range of levels involved in naming, from neural activity to overt behaviors, and determine what consequences such models have for observable behavior and neuroscience measures (e.g., Arbib, Bischoff, Fagg, & Grafton, 1995).

Behavioral dissection of the naming process and related functions has been brought to a high degree of sophistication through a number of tasks, including such methods as analysis of naming errors and prompts (Wingfield, Goodglass, & Smith, 1990), naming coupled with auditory lexical decision (Levelt et al., 1991a), and picture–word interactions (Starreveld & La Heij, 1996). Neuroscience data about naming and related functions is now accumulating through a large assortment of investigative techniques in humans, including direct cortical electrical stimulation (Gordon et al., 1991; Malow et al., 1996; Ojemann, Ojemann, Lettich, & Berger, 1989); transcranial magnetic stimulation (cf. Coslett & Monsul, 1994); thalamic stimulation (Ojemann, 1975); regional cerebral blood flow measured by positron emission tomography (PET) (Bookheimer, Zeffiro, Blaxton, Gailard, & Theodore, 1995; Damasio, Grabowski, Tranel, Hichwa, & Damasio, 1996; Martin, Haxby, Lalonde, Wiggs, & Ungerleider, 1995; Martin, Gagnon, Schwartz, Dell, & Saffran, 1996) and by functional magnetic resonance imaging (MRI) (Binder et al., 1996); regional metabolic activity via fluordeoxyglucose (FDG) PET; regional electrical activity recorded from the scalp, including evoked potentials (van Turennout, Hagoort, & Brown, in press); regional electrical activity recorded directly from the cortex (Crone et al., 1994; Ojemann, Fried, & Lettich, 1989); magnetoencephalographic recording (Salmelin et al., 1994); and even some single-unit data (Ojemann, Creutzfeld, Lettich, & Haglund, 1988). Multicell recordings and high-resolution (~ 0.5 mm) optical imaging are now available from animals performing functions related to those used by naming, such as face recognition (e.g., Wang, Tanaka, & Tanifuji, 1996). These various neuroscience methods, coupled with appropriate behavioral task sophistication and interpretative cautions (Sarter, Berntson, & Cacioppo, 1996), permit a form of direct access to the internal stages and processes involved in naming that complements the internal access available through behavioral methods.

There are excellent reasons to believe that the processes and neural mechanisms that underlie the cognitive operations involved in naming will prove to be similar to those used for other cognitive operations, because of the inherent conservatism of the evolutionary processes that built them (see Butler & Hodos, 1996, esp. p. 473; Kaas, 1993). The lessons learned from an understanding of naming therefore should be broadly applicable to other attempts to relate cognition to the brain.

What, then, has been learned? Theories of visual confrontation naming have been discussed in a number of recent papers (e.g., Dell et al., submitted; Glaser, 1992; Goodglass, 1993; Hillis & Caramazza, 1995; Humphreys, Riddoch, & Quinlan, 1988; Johnson, Paivio, & Clark, 1996; Kosslyn & Chabris, 1990; Levelt et al., 1991a; Theios & Amrhein, 1989; Tippett & Farah, 1994). There have also been several recent reviews of naming disorders (Goodglass, 1993; Henderson, 1995) and of the neuroanatomic underpinnings of naming processes (Goodglass, 1993; Henderson, 1995; Tranel & Damasio, chap. 3, this volume).

The goal of this chapter is somewhat different. It aims to contribute what might be termed an *empirical axiomatic* approach towards constructing and validating theories in cognitive neuroscience (for related approaches, see e.g., McClelland, 1993; Stone & Van Orden, 1994). The hope is that the field can avoid the situation McNamara (McNamara, 1992) has lamented occurs in cognitive psychology: "Theories in cognitive psychology are not based on natural law or on known biological mechanisms. Consequently, all theories are ad hoc. . . . theory testing is usually a war of attrition" (p. 658). This need not be true in cognitive neuroscience. Many different sources of evidence are allowed (and even welcome); theories should be constructed accordingly. There are two main parts to the empirical axiomatic approach.

The first part is to isolate what may be termed the axioms that have been used in theories of naming, and to place them within a common framework, so that they can be clearly identified. The second part of the empirical axiomatic approach is to actually identify the empirical underpinnings of the axioms: what assumptions are based on what evidence, and not just the evidence that they produce a working model!

The reasons to aim for an empirical axiomatic approach are rooted in both the fundamental problem of understanding mental operations and their neural substrates, and in the practical problems of evaluating competing theories. Naming is, without a doubt, like so many other mental processes in that it must be a multistage, dynamic process. Multistage dynamic processes are almost certain to have extremely complex behaviors that are very sensitive to the properties of their internal processes. So assumptions about any components of a model are likely to have radical consequences in terms of its overt behavior. Conversely, it is often difficult if not impossible to infer the internal mechanisms of a process from its overt performance (cf. Uttal, 1990). It would therefore seem wise to stay close to the edge of Occam's razor, and accept only the most proven building blocks for theories of the naming process.

Experimentally, the empirical axiomatic approach is an attempt to correct a serious problem, brought on by the proliferation of theories that work only too well. Most existing theories are presented as entireties, often with a few extra assumptions added in the general discussion. This makes them capable of accounting for the data, but conflates and camouflages many different theoretical and empirical issues. Although unpacking and rearranging the theories may be a bit procrustean, it is the only honest way to compare their actual mechanisms (cf. Stone & Van Orden, 1994).

The structure adopted for this review will be as follows. We will first give more precise definition of the task, and of some of the factors influencing task performance. Next, the elements that have been found useful in theories of naming will be discussed. The discussion will move from large-scale to small: from the major processing subcomponents thought to be involved in naming, to their postulated interconnections, to the supposed internal workings of these processing stages, and then to the possible neural underpinnings of these within-stage processing abilities. Literature citations will generally be to the most recent publications that can serve as links to earlier studies.

It is hoped that what results is not so much a potpourri of theor*ies* of naming, but a convergence on the outline of a theor*y* of naming that is most supported by the available evidence and theoretical reasoning.

THE NATURE OF THE TASK

The visual confrontation naming task that is the focus of this review consists of the presentation of an object or picture to the subject, with the task being to name the object as quickly and as accurately as possible. Several standardized sets of pictures have been developed, most notably the Boston Naming Test (Goodglass, Kaplan, & Weintraub, 1983a; Borod, Goodglass & Kaplan, 1980), the Peabody Picture Vocabulary Test—Revised (Dunn & Dunn, 1981), and the drawings normed by Snodgrass and Vanderwart (Snodgrass & Vanderwart, 1980).

Response Latencies

For adults naming the more frequent items on these types of tests, latencies of correct responses (from picture onset to onset of the spoken response) typically range from 400 to 1500 msec (Goodglass, Theurkauf, & Wingfield, 1984). This is not to say that individuals cannot normally have longer latencies. It is, of course, an everyday experience that for some objects, and particularly for the names of people, naming latency can be measured in minutes, or sometimes hours or days. However, responses with latencies longer than about 1500 msec in normal subjects probably represent a higher proportion of trials where there is conscious mental search and reprobing for the name (Goodglass et al., 1984). At the very least, these longer-latency responses cannot be presumed to be generated by the same direct mechanisms that give rise to the faster responses.

Subjects who are not normal—such as fluent aphasics being studied for their naming errors (e.g., Martin, Wiggs, Lalonde, & Mack, 1994; Dell et al., submitted) or patients with semantic deficits of various kinds (e.g., Hart & Gordon, 1992)—pose an interesting problem in this regard. Such subjects would also be expected to show a pattern of more rapid responses that are the direct product of their available naming mechanisms, and slower responses that have more convoluted origins. Unfortunately, there is no easy way to determine which of their responses are informative and which are adventitious. Latency measurements would be a good first attempt, but latencies or latency distributions are rarely measured or reported. As their naming latencies are often greatly prolonged in general, naming data from these subjects has to be interpreted with extreme caution.

Factors Influencing Naming Accuracy and Latency

Even for accurate, rapid responses, a large number of variables have been shown to influence latency and accuracy of visual confrontation naming. It is not yet clear which of these variables are more primary than the others. This is partly because many of these variables are highly intercorrelated, and partly because most studies have only examined subsets of the relevant variables, and not necessarily under all the conditions that might be expected to show their influence. Also, many reported studies have simply lacked the statistical power to either rule in or rule out a role for any particular variable.

As expected from even an intuitive understanding of the naming task, variables of influence can be roughly grouped into visual variables, variables related to object properties, and variables related to the name itself.

Visual Qualities

Factors that putatively affect the ease with which the external visual stimulus makes contact with the internal visual representation of the object quite understandably influence the accuracy and speed of naming (for reviews, see Johnson et al., 1996; Kosslyn & Chabris, 1990). Visual area and visual angle influence encoding speed (Snodgrass & McCullough, 1986; Theios & Amrhein, 1989). The visual complexity of the stimulus can affect naming speed and accuracy (Berman, Friedman, Hamberger, & Snodgrass, 1989). The visual realism of the stimulus can also affect naming. Subjects who are illiterate or semiliterate name real objects more accurately than color pictures, and color pictures better than black-and-white line drawings (Reis, Guerreiro, & Castro-Caldas, 1994). Educational status would be expected to affect naming accuracy and latency in at least two ways: Lack of visual familiarity would impose an extra processing burden on the input. Less familiarity at every level of processing would slow throughput throughout, and magnify inaccuracies or delays at the early stages. Therefore, it is not surprising that lower educational level can expose the different demands that different types of stimuli make on initial processing. It is likely that damage and inefficiency in the subsequent processes involved in naming, as may occur after brain injury, can also unmask visual quality effects, but no studies have specifically addressed this issue.

Semantic factors such as imageability, concreteness, prototypicality (Morrison, Ellis, & Quinlan, 1992), semantic category (Morrison et al., 1992), and operativity (Nickels & Howard, 1995) have been shown to influence naming in either normal subjects, in patients with aphasia, or in both groups (Nickels & Howard, 1995).

Characteristics of the *name* itself—such as the degree of agreement about the name for the item, its length in syllables, and, most importantly, its frequency or familiarity—have been found to influence naming latency and accuracy in normal subjects and in patients with naming impairments (see Johnson et al., 1996, for a review).

Frequency, mainly but not exclusively frequency of the name, has proven to be a very potent variable. At least four interdigitated issues have to be disentangled. One is whether frequency or some other highly correlated variable, such as age of acquisition, is the critical variable. Some recent studies have indicated that *age of acquisition*—either rated age, or more objective estimates of the age at which the word was learned—may be as important an influence as frequency. Group studies have shown it to be an important factor (Rochford & Williams, 1962; Feyereisen, Van Der Borght, & Seron, 1988), as have individual case studies (Hirsh & Ellis, 1994; Hirsh & Funnell, 1995). Age of acquisition accounted for essentially all of the effects previously attributed to word frequency in other studies of naming (Morrison, Ellis, & Quinlan, 1992; Nickels & Howard, 1995). Age of acquisition has also been reported to be more important in determining the speed of reading words out loud than is word frequency (Morrison & Ellis, 1995). Whether age of acquisition or some other variable will supplant frequency in models of the naming process cannot yet be determined, but fortunately theories of the microprocesses involved in naming can accommodate a wide variety of related variables, as will be seen.

Keeping frequency as the primary concern for now, there is still the issue as to how it should be measured. Measurement of frequency by objective counts is common, although these are often inaccurate for less common items or individual subjects (Breland, 1996; Gernsbacher, 1984; Gordon, 1983). Subjective ratings of frequency have often been apparently more accurate in other domains (Gernsbacher, 1984; Gordon, 1983), and have been found to be useful in understanding the results of some naming studies (Hart et al., in preparation). However, subjective frequency estimates may incorporate many different variables besides frequency as it is commonly understood.

Across the situations under which naming is usually tested, with items for which there is fairly good name agreement, presented in a standard-ized visual format, the frequency that has the greatest effect on naming has been that of the object name (for review, see Johnson et al., 1996). Con-ceivably, the frequency of the object as a visual experience, the frequency of the semantic category, and even the frequency of its component sylla-bles could influence naming, but in practice their effects are negligible. Conversely, the frequency of the name is not related to visual identification processes (e.g., Jescheniak & Levelt, 1994), nor to name articulation (Jesche-niak & Levelt, 1994; Wheeldon & Monsell, 1992). The frequency of the spo-ken word (Howes, 1966), as opposed to the frequency of the word in print, might be expected to be an even better correlate of naming, but this com-parison has not been made.

THE PSYCHOLOGICAL AND NEURAL COMPONENTS OF NAMING

Stage Analysis—General considerations

The search for modularity—that is, a decomposition into functionally in-dependent subprocesses concerned with different codes or operations—has been accepted as a guiding principle of cognitive science and cognitive neuroscience. It has been suggested that all complex systems actually *re-quire* a parcellation of tasks into subprocesses (Simon, 1962; see also Raff, 1996, pp. 325ff; Riedl, 1978). More recently, neural network theory has pro-vided a more detailed justification for this, suggesting that tasks ultimate-ly become decomposed into subprocesses through dynamic competition in connectionist architectures (Jacobs, Jordan, & Barto, 1991; Polk & Farah, 1995). The logistics of neuronal interconnection may also predict fragmen-tation of neuronal processing into subsystems (Ringo, 1991).

Two-Stage Characterizations

In the case of visual confrontation naming, the minimum decomposition that appears to be absolutely necessary is into two parts, vision and lan-guage. These two major systems are clearly completely independent of each other, both across the animal kingdom and across individual cases.

Three-Stage Characterizations

Most investigators in the area have accepted three stages as the minimal set involved in naming: a visual object-recognition stage; a semantic stage;

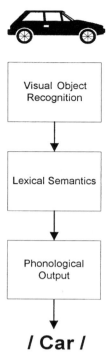

Figure 1   The basic three-stage model of visual confrontation naming. Information flow is feed-forwards. An intermediate, amodal semantic stage is obligatory. (Figure ©1996 Intelligence Amplification,
Inc. Reprinted with permission.)

and a phonologic output stage (see Figure 1). The visual object-recognition
stage is presumed to identify the object visually, but is often assumed to be
ignorant of the uses and associations of the object. That knowledge is reserved for the semantic stage (perhaps more properly, a lexical semantic
stage). The semantic stage, although the repository of all nonvisual information about the object (and also perhaps, lexically encoded visual information), is ignorant of the phonologic form of its lexical labels. That phonologic knowledge is within the phonologic output system.

Many accounts of naming and of its disorders have found this three-
stage model, with the middle stage being an amodal semantic stage, to be
sufficient. Indeed, a great many reported studies have been concerned with
establishing how their data fit various aspects of a three-stage model. For
example, in different patients and patient groups, errors in visual-confrontation naming have been attributed to the visual-input stage (Kirshner,

Casey, Kelly, & Webb, 1987; Tweedy & Schulman, 1982), the semantic stage (e.g., Hart, Berndt, & Caramazza, 1985; Hart & Gordon, 1992), or the phonologic output stage (Howard, 1995; Kay & Ellis, 1987).

Despite the explanatory success of the three-stage model, it may not always be possible to distinguish its consequences from those of a two-stage model with modality-specific input routes that have direct connections to phonologic output, as Johnson, Paivio, and Clark (1996) have emphasized. For example, the naming deficits in aphasia are generally multimodal, and this has been taken as one line of evidence for the existence of a central, amodal semantic deficit (e.g., Goodglass, Barton, & Kaplan, 1968). However, a disconnection of multiple inputs to a phonologic output system could give similar results. Of course, the functional lesion in the latter case would not necessarily account for the other semantic deficits these patients can show (e.g., Goodglass & Baker, 1976).

The debate between the models with an amodal semantic system (nearly all existing models) and those without (Johnson et al., 1996) can be recast as a debate about the importance of two different pathways in naming: one, involving direct sensory system-to-phonologic output connections; the other, with an intervening semantic stage. Because even some orthodox models of naming (e.g., Goodglass, 1993) accept the possibility of direct sensory-to-phonologic connections (as will be noted below), the issue seems very hard to resolve with current experimental tools.

Postsemantic Substages

Many theories of naming find the assumption that semantic representations are directly mapped onto whole-word phonologic representations to be perfectly adequate, as Starreveld and La Heij (Starreveld & La Heij, 1996) have pointed out. However, in order to account for more general characteristics of word retrieval, a finer parcellation of the processes involved in word selection and phonologic production has been necessary; see, for example, the proposals by Levelt and his colleagues (Levelt, 1989; Levelt et al., 1991a), Garrett (Garrett, 1980), Roelofs (1992), and by Dell and his colleagues (Dell, 1986; Dell, 1988; Dell, 1990; Dell & O'Seaghdha, 1991, 1992). An outline of the resulting more detailed stage model is depicted on the right-hand side of Figure 2.

In this more finely divided model, *visual-feature identification* and then *visual-object identification* are still the initial stages. (Note that although they are depicted here as being sequential, visual-object identification may in fact go in on parallel with visual-feature identification, with the processing sequence being more for different *degrees* [coarse to fine], rather than for different *types*.). The next stage is *lexical semantic*, involving the retrieval

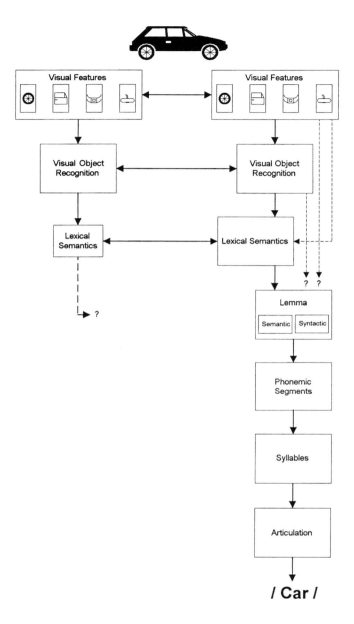

Figure 2 The elaborated model of visual confrontation naming showing more details of the processing required for selection of word phonology, the processing accomplished by the nondominant hemisphere (in most individuals), and possible direct connections (as dotted lines: - - - - -). Information flow is feed-forwards; no feedback loops are yet justified by the evidence (except for external auditory feedback, not shown). Information flow is successive, and probably partial and in cascade; see text. (Figure ©1996 Intelligence Amplification, Inc. Reprinted with permission.)

of lexical concepts. Many lexical concepts may apply to the same picture. For example, the picture of a car may justifiably elicit the lexical concepts [Ford] and [hatchback]. These concepts next activate a *lemma*, representing the independent semantic and syntactic specifications of what can be regarded as a protoword (but not its phonological properties). The existence of lemmas intervening between semantics and output phonology can be justified by the need to accommodate a wide variety of phenomena (Kempen & Huijbers, 1983; Levelt, 1989; Roelofs, 1992). It can also be justified by the logical need to map between semantic representations and phonological representations that are both distributed over sets of their respective nodes. An intermediate level is necessary to solve the exclusive-OR problem this poses (Dell et al., in press).

The functioning of this interface between lexical semantics and phonologic production has been closely examined. Levelt (1989) has argued that only one lemma is selected. However, there is evidence that many phonologic word forms must be activated (Laine & Martin, 1996), a conclusion that on the surface is incompatible with the single-lemma hypothesis, and which is consistent with the general assumption that multiple representations are contacted at each stage of processing (see below).

Regardless of whether one lemma or many are coactivated, the lemma is what specifies the protoword's *phonemic segments* and *metrical structure*. Next, the *syllables* corresponding to groups of segments are generated. Finally, in the *articulatory* stage, instructions are given to the motor systems of the lips, jaw, tongue, pharynx, larynx, and respiratory systems to orchestrate the actual production of the speech sounds. This is, of course, still just a simplified account of postsemantic lexical access. There is widespread agreement about even further details (such as the existence of *phonological frames;* see Dell et al., in press), and widespread disagreement about many others.

The available neurologic evidence from humans does suggest that it is necessary to make one important modification to the psychological model of the stages involved in visual confrontation naming; that is, to recognize the existence of processing in the nondominant hemisphere (see Figure 2). There seems to be universal agreement that the human cerebral hemispheres have two anatomically distinct systems for visual object recognition. It is even possible that the nondominant (usually the right) hemisphere is more adept at visual-object recognition than is the left (Sergent, 1987). Evidence that lexical semantics—for lexical concept retrieval—is also possible in the right hemisphere has been more controversial. However, enough evidence has accumulated for right-hemisphere participation in lexical semantics (Chiarello, 1991; Coney & Abernethy, 1994; Hadar, Ticehurst, & Wade, 1991) as well as for other verbal functions (e.g., Berthier et al., 1991; Patterson, Vargha-Khadem, & Polkey, 1989) to indicate that the following is a reasonable characterization of the lexical semantic capacities of the right hemisphere: Its capabilities vary across individuals, but on the average, the nondominant hemisphere possesses some lexical semantic capabilities. These generally appear to be more sketchy, more concrete, and less forceful (that is, activated more slowly and to a lower ultimate degree than the lexical semantic representations of the dominant hemisphere). Therefore, the nondominant hemisphere's contribution to lexical semantics and visual confrontation naming is often obscured by the left hemisphere's contribution, or only shows up as latency differences on lateralized naming (McKeever, Seitz, Krutsch, & Van Eys, 1995). It would rarely if ever be detectable after right brain injury, as is indeed the case (Glosser & Goodglass, 1991). However, when left brain injury puts more of the burden on the nondominant hemisphere, its capabilities are more clearly exposed, and might even expand somewhat.

Whether the nondominant hemisphere is ever normally capable of independent phonologic production, and at what level (e.g., phonetic or syllabic) is even more debated, as depicted in Figure 2. There is good evidence, however, that the nondominant hemisphere can produce words in auditory repetition after dominant-hemisphere stroke (Berthier et al., 1991; Ohyama et al., 1996).

NEUROANATOMIC LOCALIZATION

Despite the duality of processing routes available for naming (at least, for its initial stages), a number of lines of evidence point to the important conclusion that the processes and routes involved in naming are relatively discrete. Lesions in many areas of the brain do not appreciably affect naming; discrete lesions in some other regions drastically do. Imaging data, such as that shown in Figure 3, also supports the notion that the psychological stages identified as being responsible for naming correspond to relatively discrete neuroanatomic regions. The neuroanatomic correlates of naming and its impairments are discussed more completely in Goodglass (1993) and in Tranel et al. (chap. 3, this volume). The conclusion important for the present purposes is that the stages involved in naming are the product of relatively circumscribed cerebral regions. There are, in fact, now good theoretical reasons to expect neuroanatomic clustering of the "same" functions, and to disavow an agnostic, mass-action view of cerebral localization. Theory and simulation suggest that even when related functions are initially separated, several pressures will force them into contiguity: the need for rapid transmission and for economy of wiring, for example (cf. Jacobs & Jordan, 1992).

A combination of the results of lesion studies (e.g., Hart & Gordon, 1990) and activation techniques such as O^{15} PET (Bookheimer et al., 1995; Martin

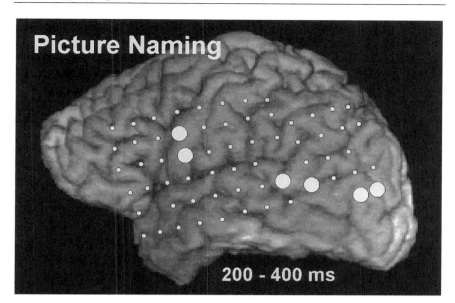

Figure 3 Relative focality of regions engaged by visual confrontation naming. Here, it is illustrated by results of direct cortical electrical recording during visual-confrontation naming in a single subject. Electrocortical activity recorded from an indwelling subdural electrode array, averaged over approximately 70 trials. Electrodes showing significant reduction in alpha band power (taken as an index of regional cerebral involvement) are indicated as large dots. At this point in time, 200–400 msec after onset of the stimuli, focal activation is apparent in the left (dominant) lateral occipital lobe, the left posterior superior temporal lobe, and in the left sensorimotor and inferior frontal regions. (Crone, Gordon, Hart, Lesser, unpublished data; see Crone et al., 1994) (Figure ©1996 N. Crone, B. Gordon, J. Hart, & R. Lesser. Reprinted with permission.)

et al., 1995, 1996) and direct cortical recording (Crone et al., 1994; see Figure 3) suggests that the stages corresponding to lexical-semantics and phonologic output processing may depend upon cortical regions on the order of 2–3 cm in linear extent. (Note that size estimates from the lesion data have to be adjusted, because lesions producing persisting deficits are probably of greater extent than the territory that is normally critical for the function; cf. Selnes, Risse, Rubens, & Levy, 1982; Gordon, Hart, Lesser, & Selnes, 1994).

FURTHER DECOMPOSITION AND ITS LIMITS

The parcellations schematized in Figure 2 cannot be the smallest quanta of processing involved in naming, nor are the neuroanatomic regions discussed above necessarily the smallest neural building blocks.

Visual-object identification, because it can be studied in nonhuman primates and other animals in far more detail than is possible in humans, is perhaps one indicator of what may be found when subsequent stages of the object-naming process are examined in a similar fashion. Visual-object identification is known to be the product of initial identification of visual features (such as shape, color, and motion), and a recognition of the configuration of these features as unique objects, such as a toy or a face (Tanaka, 1993; Maunsell, 1995; Lueschow, Miller, & Desimone, 1994). The fine-grained neuroanatomy and single-unit neurophysiology of visual-object recognition are now being mapped with optical recording (with a resolution of < 0.5 mm; see Wang et al., 1996) and by simultaneous microelectrode arrays (Lueschow et al., 1994). What is perhaps the most striking conclusion from these studies, for our purposes, is that very fine featural distinctions have been identified, confined to very small regions of cortex (~ 0.5 mm).

It is possible, however, that the data from visual-object identification may overestimate the psychological and neural elaboration that will be found for subsequent stages involved in naming. Vision has been such an important function for primates, and it has such a long evolutionary history, that it is the responsibility of a very large fraction of the total cortical area of the primate brain (Van Essen, 1979). The cortex responsible for vision may represent an atypical example of extreme functional specialization and segmentation.

It is certain, in fact, that the quest for behavioral modularity and a corresponding neuroanatomic modularity cannot be endlessly rewarded in the same ways they have been in the past. There are now indications that the cortical parcellations that have been part of the support for functional subdivisions are actually less distinct, and far more volatile, than was suspected (Singer, 1995; Swindale, 1990; Wang, Merzenich, Sameshima, & Jenkins, 1995). Also, detailed studies of the neural activity within individual brain regions, at different spatial and temporal scales, has in some cases not shown an orderly spatial segregation of different functions. Instead, different functions may be accomplished by the same cortical region, kept distinct only by temporal and other, nonspatial codes, which are as yet only dimly appreciated (e.g., Bullock et al., 1995; Singer, 1993), and which are necessary to solve the related problem of binding different representations together (von der Malsburg, 1995).

BETWEEN-STAGE INTERACTIONS

Moving backwards to a more macroscopic level where it is still reasonable to talk about different stages (and different regions) as being responsible for naming, the next logical question is then how these stages interact with each other. There are at least four questions that must be answered about the relationships between stages:

A. Which stages are forwards connected to which other stages?
B. Which stages are backwards connected to other stages (feedback connections)?
C. What is the temporal sequence of the connections between stages?
D. What is the nature of the information that is conveyed between interconnected stages?

Forwards Stage Interconnections

There is widespread agreement that the stages in Figure 2 have forwards connections, as already indicated in the diagram; that is, information from a stage is passed forwards to the next stage in line.

But one controversial issue about feed-forwards connections in naming is how direct they can be. Can there be direct connections between the representation of a visual object and its phonological form, without semantics intervening (an issue alluded to earlier)? Between visual features and the phonologic form? The evidence for and against these possibilities can be summarized in the following way: It seems likely that at least some objects can have direct connections to phonologic word forms, without semantic intervention, simply as a consequence of learning in early childhood or at other times when no semantic information has been given. One can even imagine a distorted educational system wherein naming is taught independently of understanding of what is being named. (Some medical training may qualify). So the existence of direct connections between visual-object representations and their phonological labels may be true for some objects across a large number of people, and for some people with respect to a more idiosyncratic set of objects and names. Analogous direct visual "object" to phonologic name links are widely accepted (although not universally so) in the literature on reading, where one class of theories concludes that short, frequent word forms in particular are directly linked with their phonologic translations (Besner, in press; Buchanan & Besner, 1993; Coltheart, Curtis, Atkins, & Haller, 1993).

The evidence from brain pathology is suggestive of direct connections, but far from conclusive. Some cases have been described in which visual confrontation naming is preserved despite profound deficits in semantic comprehension (Brennen, David, Fluchaire, & Pellat, 1996; Heilman, Tucker, & Valenstein, 1976; Kremin, 1986; Shuren, Geldmacher, & Heilman, 1993), which could be interpreted as evidence of a preserved direct route despite damage to the semantic system. However, as Brennen et al. (1996) pointed out, there is a problem with trying to use such evidence against a requirement for semantic mediation. Markedly impaired semantic comprehension and intact naming could occur in semantically mediated models, if the amount of activation needed for generating the name were far less than that required for comprehension. Therefore, a dissociation between the two because of pathology may be due to the different demands of the tasks, rather than to the existence of an entirely separate route.

On the other hand, the direct route hypothesis is also difficult to exclude. Both psychologic (e.g., Schriefers, Meyer, & Levelt, 1990; Goodglass et al., 1991) and electrophysiologic (e.g., van Turennout, Hagoort, & Brown, in press) data strongly suggest that semantic access occurs prior to phonologic access in naming. But none of these studies has reported the by-subjects and by-items analyses—of sufficient statistical power—that would be required to test the most reasonable form of the direct route hypothesis, which is that the effect only occurs for some items, and perhaps is most discernible in only some subjects. Therefore, direct routes have been included as possibilities in Figure 2, but not as certainties.

Backwards Connections

One of the most controversial parts of the architecture of some proposed theories of naming (and of word retrieval in general) is whether there are backwards connections between the stages. The basic issue is whether information from what appear to be logically subsequent levels of processing can influence earlier stages of processing. Forwards activation only is assumed in Levelt et al.'s (1991a) model of picture naming, and in Humphreys et al. (1988) model. Backwards connections are adopted by Dell's model of naming and of word production in general (Dell, 1986, 1988, 1990). Backwards connections are present between every stage in Tippett and Farah's (1994) computational model of naming. Backwards connections have also been invoked in explanations of related processes such as word naming (reading outloud) as in Seidenberg and McClelland's (1989) model and Plaut et al.'s (1996) extension of that model.

The question of backward connections is closely coupled with the question of exactly which stages are considered to be interconnected, and with the question of the time course of processing between stages. Backwards connections that allow feedback can also be the cause of properties that are of central importance for some models, such as the emergence of attractor dynamics in some connectionist models of reading (e.g., Hinton & Shallice, 1991; Plaut et al., 1996). However, it is important to resist the postulation of backwards connections simply for convenience, and consider only the actual evidence that they exist.

Although many pairs of stages have been variously assumed to be coupled through feedback, a hypothetical feedback between phonologic word access and the preceding lexical access stage has been the target of most active debate. Some aspects of this debate, as seen in the published interchange between Dell and O'Seaghdha (1991) and Levelt et al. (Levelt et al., 1991b), are instructive for the general question of whether feedback between stages exists, and how it should be tested. The debate has been complicated by the usual difficulty in actually isolating the issue to be tested. Dell's model, for example, not only incorporates backwards connections, but also spreading activation, and a temporal cascade of processing (if for no other reason than as a result of the backwards connections!).

But the fundamental problem is that most existing psychologic and neuropsychologic methods do not have the resolving power necessary to settle this debate. Whether additional neurophysiologic methods will be sufficient remains to be determined (see Hendriks & McQueen, 1996, pg. 34ff). So far, nothing stops the existence of feedback between processing stages, but no evidence yet definitely requires it, either.

SAME-STAGE FEEDBACK

Note that there is another logical possibility that is occasionally invoked in models of related processes (such as reading; see Plaut et al., 1996), which is that output from a stage feeds back upon itself. No model of naming has yet suggested this occurs. Yet the neuroanatomic evidence for within-region (presumably, within-stage) feedback is overwhelming (Shepard, 1990; see Ullman, 1995, for specific use of intracortical feedback). Self-feedback allows an enormous number of new processing dynamics to appear, including attractor dynamics. These have immediate utility for models of cognitive function (see Neural Mechanisms section, below). Although not included in the schematic model of Figure 2, within-stage feedback with its attendant dynamics is almost certain to occur at some neurophysiologic level in the stages involved in naming, as we will note later.

The Temporal Sequence of Processing between Stages and the Amount of Information Transmitted

The temporal flow of processing, and the amount of processing done by a stage before information is passed along or used, are two very contentious issues in cognitive science (for recent reviews, see Liu, 1996; Meyer, Irwin, Osman, & Kounios, 1988; Miller, 1990). They are logically distinct (as pointed out by Liu and others), but they have often been comingled. For example, the cascade model (McClelland, 1979) posits that not only are stages temporally overlapping, but also that partial information is available from one stage for use by the next. Because this conflation has been adopted by many models of naming, it will be impossible to completely disentangle.

With respect to the temporal sequence of processing, there are two major positions that have been taken. One is that processing of a prior stage in some sense ends before processing of a subsequent stage begins. This is a traditional assumption (Donders, 1868/1969) and the one used by Levelt et al.'s (1991a) model of naming and word retrieval. The experimental utility of the Levelt et al. model shows that it is still a very reasonable assumption in many cases.

A contrasting assumption is that the temporal sequence of processing between stages is overlapping. This is the assumption made in the models proposed by Dell (e.g., Dell & O'Seaghdha, 1992), Humphreys et al. (Humphreys, Riddoch, & Quinlan, 1988), and many other recent models of naming in normals, and in the accounts by Martin, Dell, and their colleagues of the errors in naming that result after brain injury (Martin et al., 1994; Dell et al., in press). (Note that these models also explicitly assume that partial information is available from one stage to the next).

Although the strictly serial model of Levelt et al. may be logically defensible, and although it clearly can fit most of the data (if not all), models with temporal overlap and partial information flow are strongly supported by a number of different lines of more direct experimental evidence. Psychophysical and evoked potential measures (Miller & Hackley, 1992; Smid, Mulder, Bocker, Van Touw, & Brunia, 1996) are highly consistent with cascade processing, at least for some cognitive functions. Neurophysiologic data can also be construed this way (keeping in mind, of course, that neurophysiologic data cannot yet be brought into correspondence with most of the cognitive constructs being considered here). Processing to a high degree of visual identification can be extremely rapid (Thorpe, Fize, & Marlot, 1996). Single-unit recordings along the pathways responsible for visual-object recognition appears to show a very rapid spread of information from initial stages to what are presumably subsequent stages of object processing (e.g., Desimone, 1991). And there is some evidence from our own work with direct cortical recording during naming that regions that may correspond to stages are activated in overlapping, rapid sequence (Crone et al., 1994; see Figure 3). Therefore, it appears reasonable to incorporate temporal overlap and partial information flow into psychological models of the naming process.

The Nature of the Information Conveyed between Interconnected Stages

There now seems to be widespread agreement—at least in the psycholinguistic literature—that the connections between stages convey highly detailed information, despite the narrow line drawn to indicate their presence (as in Figures 1 and 2). How many of the details of processing within a stage are conveyed to its following stage is a hidden assumption that can, in fact, have important consequences in certain situations. The extreme assumption is that there is only a single channel or pipeline between stages. This seems to have been the implicit assumption in the neurologic literature for many years (Gordon, 1982). It was assumed that brain damage might selectively destroy a channel, but it never seemed necessary to assume that the connections between stages were such that some types of communication could be destroyed, although others were left intact. This notion is no longer either plausible nor supported by the evidence. But what is widely accepted is the notion that processing stages are opaque; only their outputs are seen by subsequent stages. Whether processing stages are truly opaque, or whether they are translucent to a greater or lesser degree, remains to be seen.

WITHIN-STAGE PROCESSING

Representation within a Stage or Code

How specific information is actually represented and processed within a stage is the motive force for the global processing of naming, and therefore in many ways the most critical assumption of all. Given all the controversy about the rest of the machinery, there is nearly universal concurrence (in keeping with the rest of cognitive science) that within each stage or representational level, individual components are represented by distinct *units* or *nodes*.

What level is being considered does not matter. At the initial input stage, different units correspond to different objects (in a prototypical, abstract visual sense). At the lexical-semantic stage, different units may correspond to different semantic features. Lemmas are units (or actually, may be made up of separate semantic and syntactic units). Lexemes, phonemes, and any other item under consideration have always been assigned to units or nodes. This has been true regardless of the controversies over what is exactly in a level of representation; whatever type of items a naming theorist endorses, they are represented as units or nodes.

Nodes are not inactive atoms or quanta; as in chemistry or physics, they have been postulated to have properties that give rise to the observed dynamics of naming. Abstracting from a number of sources (but cf. McClelland, 1993), it is possible to list a set of core postulates from which various theorists have drawn (more or less completely):

1. Nodes correspond to the *items* that exist—in the experience of the particular subject—*at that level of coding*. Presumably, nodes get created by exposure to new items. (They must start out with very little activation level [as defined below], of course, because the new item by definition has a low frequency on its first exposure.)

2. Nodes get input *in parallel* from prior stages. That is, the information generated by the preceding stage is broadcast to all the nodes in the subsequent stage.

3. Nodes within a level are *independent;* what happens to one node does not necessarily affect another. Activation of other nodes within the level can be altered only if (a) they are explicitly interconnected or (b) if the theorist proposes a specific rule governing their mass activity, as listed below.

4. Each node has a unidimensional property, termed its *activation* level.

a. The activation level of a node can vary over time.

b. Each node has a baseline activation value, the one existing before the start of the experimental trial (or experiment).

c. In most accounts, the level of this baseline activation is related to the frequency of the item that that node represents. The greater the frequency, the higher the resting activation level of that node. (Equivalently, frequency may be represented by a change in a firing threshold, discussed below.)

d. Input from the prior stage can add activation to a node. The amount of added activation depends upon how closely the input matches what the node represents.

e. The *dynamics* of activation are generally assumed to be noninstantaneous. Activation in response to even a single pulse input builds up over time. However, this buildup does not go on indefinitely. For many reasons, including avoidance of "heat death" and network stability, it is often assumed that the growth of activity in a node obeys a sigmoid curve: slow at low levels of input, faster at higher levels, and saturating at still higher levels (see Figure 4).

f. Once activated above its baseline level, the activation level of a node spontaneously decays back to baseline, over a finite period of time (see Figure 4).

g. Although perhaps not part of the classical conception of nodes, activation levels may have *random fluctuation* (random noise) (McClelland, 1993). The magnitude of this random noise may be positively correlated with the node's activation level (cf. Dell et al., submitted).

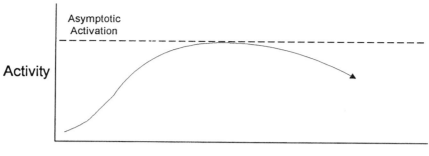

Figure 4 Activation function for a single unit, at a single level. (Figure ©1996 Intelligence Amplification, Inc. Reprinted with permission.)

5. If a node is "sufficiently" activated, it

a. then reaches a "threshold" and then sends activation on to a subsequent stage (can broadcast to a subsequent stage);

b. *can* inhibit all other nodes at that level (see the suggestion made for semantic-level nodes by Nickels and Howard (1994).

c. may get additional activation added, as a bonus for reaching a certain level of activation (as Dell has suggested occurs at the lexical level). (As Nickels & Howard [Nickels & Howard, 1994, p. 314] pointed out, this is equivalent to inhibition of other nodes at that level.)

Note that (5a) and (5b) are both mechanisms that ensure a winner-take-all allocation of activation across nodes within the stage.

6. The threshold setting may be

a. fixed in advance, or

b. it may vary with frequency of exposure, or

c. it may be subject to strategic control. As an example, if a stimulus is presented that does not excite existing nodes strongly enough to fire, the threshold might be lowered for the next stimulus (e.g., Morton & Patterson, 1980).

Triesman (Triesman, 1960) and Morton (Morton, 1964) were perhaps the first to crystalize the notion of such units and discuss their properties in print. It is not clear to this author what evidence or influence led to the postulation of these units and their properties. The notion of units seems to lend itself naturally to the world of letters, speech sounds, words, and perhaps even meanings. Activation, decay, and the rest do seem to fit with subjective impressions about how ideas or perceptions pop into our minds. On

the other hand, these nodes and their properties also bear a strong resem-
blance to idealized neurons, such as those used by McCulloch and Pitts
(1943) or those abstracted into the perceptron model of Minsky and Papert
(1969). Of course, no one has been so rash recently as to identify the hypothetical nodes or units with actual single neurons.

Even though "connectionist" or "neural network" models of cognitive
processes have been preferred as radical replacements for the classical theories, they still typically coopt the notion of these units and their properties. The main differences have been that: (a) smaller units of cognitive
knowledge are assigned to the nodes (e.g., individual phonemes rather
than words, as in Plaut et al., 1996; or perhaps even nontransparent representations have been attributed to nodes, such as may occur in the hidden
unit layer of many models); (b) the connections between units have been
assigned weights, so they play an active role in processing; and (c) the assignment of these connection weights is done through learning, not by an
outside dictate. (One other important feature of these theories, their use of
a layer of hidden units, can be viewed in this context as the recognition of
an additional processing stage, not necessarily as a difference in the nature
of the processing that is presumed to occur within a stage.) However, although these differences have been extremely important, they should not
obscure the fact that the same old units of processing are at work inside.

Regardless of whether they appear in classical or more modern models,
these units and their properties have proven to be enormously useful in
cognitive science. They have endowed theories using them with great
heuristic and explanatory power. Dell's (1986, 1988, 1990) model of word
production is an example of a model incorporating more classical nodes,
with externally assigned values. Plaut et al.'s (1996) model of single-word
reading aloud demonstrates the power of a combination of externally specified units (the basic orthographic and phonetic representations) coupled
with a connectionist learning approach to the associations between them.

Nodes are so useful that even if they did not really exist, they would have
been invented. But it is not clear what is the direct evidence—apart from
their theoretical utility—for the existence of such cognitive quanta. Part of
the problem of proving their existence, of course, is that the assumption of
nodes is made so deep into the already-existing chain of suppositions required to interpret behavioral data that it is extremely difficult to imagine
experimental tests to directly contrast the notion of independent representational units with the notion of a deeper continuum.

Perhaps some of the strongest evidence for the existence of units at a level of behavioral processing comes from the domain of auditory speech perception. The acoustic input to speech perception is clearly a continuum, yet
the behavioral translation of the speech sound continuum is typically cat-

egorical (Kluender, 1994). Also, damage to the speech perceptual system may cause dysfunction that behaves as though it is category-specific (e.g., Saffran, Schwartz, & Marin, 1976). It is perhaps in such a more peripheral and hence transparent domain that direct evidence for or against the existence of cognitive nodes will be forthcoming.

It should be noted that Stone and Van Orden (Stone & Van Orden, 1989) have specifically questioned the independent existence of nodes or units (in semantics in particular), arguing instead for what they term a *functional unitization;* that is, the emergence of unitary behavior from processing dynamics that tie a number of subsymbolic components of representations into functional, symbolic, wholes (see also Stone & Van Orden, 1994). Functional unitization seems to be exactly what is required for the modes of processing possible in neural networks to behave as though they are units of cognitive codes, as we will now discuss.

NEURAL MECHANISMS

Examination of the neurophysiologic evidence, both direct and theoretical, suggests there are many good reasons to believe that neural processing can give rise to representations that behave exactly like the units required by psychological theories. At this level, of course, the evidence and reasoning is vastly more speculative, but still collectively fairly compelling.

First, it seems apparent that the neural representation of cognitive codes must be distributed across many neural elements. "Distributed" in the sense it is being used here means that no single processing element can be said to represent that item and that item only. The notion of "neural processing element" has been deliberately left vague. Although widely identified as neurons, it is not so clear neurophysiologically that neurons are the smallest elements of processing. Considerable processing may take place in subneuronal components such as in the dendritic tree (Zecevic, 1996). Furthermore, the actual informational method nerve cells (or nerve cell subcomponents) use to convey information is still not completely known. However, the line of reasoning we will pursue here is independent of the exact nature of the neural processing element or its code.

What this line of reasoning does require is that at least a fair number (comfortably more than one) of neural-processing units are required to make up a representation. This seems to certainly be true, at least for the codes involved in object identification, semantics, and word retrieval. In face recognition, direct neuronal recordings in monkeys have suggested that as few as 14 nerve cells could code for the discrimination of as many as 50 distinct faces (Abbott, Rolls, & Tovee, 1996). This is an extreme lower bound estimate; it only applies to the discrimination of those faces, not their distinctive representation. Even so, these data (and much other data) support the notion that neural representations are the outcome of the dynamic activity of more than one neural element (although not of unfathomably large numbers; see Rolls & Tovee, 1995).

It is therefore reasonable to believe that a given behavioral stage or coding level in naming is the product of a set of neural-processing elements. The activities of these processing elements must in some ways represent the information that is being processed at that particular stage. To describe these activities, it helps to treat each neural element as a dimension in a descriptive space. The collective activity of all the relevant elements at any one point in time can then be represented as a point in a multidimensional space.

We can readily assume that these neural-processing elements are interconnected in complex ways, and that their individual operations are perhaps equally complex. It might seem at first that because little is known specifically about how these elements work or how they are interconnected, then little could be inferred about the overall behavior of the system of elements we have focused on. But there is increasing theoretical evidence that some very general inferences can be made even about the behavior of what seems to be such a relatively unconstrained system. Neural networks, particularly those with self-feedback, tend to have certain preferred states of activity (Hertz, 1995; Wang & Blum, 1995). In particular, Kauffman (1993) has argued (in what is admittedly another context) that the patterns of activity that can actually occur in such systems does not completely fill that space of possible states. Only some patterns of activity are actually possible in such networks, because of the patterns of interconnection and interdependence that necessarily exist.

Therefore, the state space of activity within such networks is not a multidimensional solid, but instead a multidimensional set of separate points. Each point represents a state space allowed by the network; in between them is the empty space of patterns of activity that cannot occur. What is also likely in such networks is that there are *clusters* of points; that is, allowable activity states that are similar to some others (see Figure 5).

It is also true of such dynamical systems that patterns of activity that are imposed on the network from outside will ultimately have to settle into the points or clusters that are allowed by the intrinsic network dynamics. In the terminology of dynamic systems theory, the network's points are *attractors*. Each attractor has a *basin of attraction*, which is the space of activities that settle into that particular point (see Figure 5).

What should be evident at this point is that the attractors of this network,

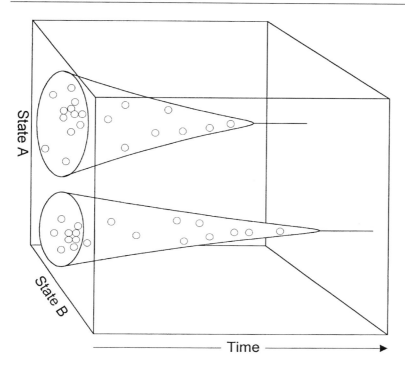

State A

State B

Time

Figure 5 The possible neural realization of units or nodes. Illustrated are two dimensions of a state space, over time. Nodes are represented as attractors, with basins of attraction. In this illustration, the upper attraction has a larger catchment area, and its activity also coalesces more rapidly. Compared to the lower attractor, the upper attractor therefore acts as though it is a node with more similarities to stimuli that might be imposed on the state space, and also has a higher frequency, with a more rapid activation rate. (Note that in general, though, size of the basin of attraction does not have to be related to the "strength" of the attractor or to the rate of convergence to that attractor.) (Figure ©1996 Intelligence Amplification, Inc. Reprinted with permission.)

and to some extent their basins of attraction, form a natural basis for exactly the unitary, "independent" nodes that are so useful as higher order elements of cognitive processing. So a combination of previously learned network connectivity, together with external patterns of input representing the stimulus, are almost certain to generate in any plausible neural network exactly the kinds of behaviors that can be used for representing distinct items. Moreover, these distinct representations would have exactly the kinds of dynamical behavior that have been thought to be necessary for such nodes: frequency sensitivity, activation, decay, even mutual inhibition and "clean-up" and "winner-take-all" behavior.

To give just two examples of how properties attributed to nodes can be explained: *Frequency effects* may correspond to the speed of convergence upon an attractor state, which in turn may depend upon how tightly the neural network has been tuned by experience. Note that in this conception, simple frequency of exposure may not be the only determinate of convergence speed. Timing of exposure, spacing of exposures, and other variables known to influence memory at the synaptic level are all likely to be codeterminants of the efficacy of connections and the tuning of the network (Kandel & Abel, 1995). As a result, it is easy to see why age-of-acquisition might be a powerful determinant of the "strength" of nodes.

Mutual inhibition, noise *"clean-up,"* and *"winner-take-all"* dynamics are often viewed as necessary and related behaviors. These could all be the result of what Reggia and his colleagues (Reggia, D'Autrechy, Sutton, & Weinrich, 1992) have termed *competitive distribution* of activity in neural networks: not the result of explicit inhibitory connections, or ad hoc additions of activation, but instead a direct consequence of a cap on overall network activity, which in turn is motivated by basic physiologic requirements to prevent runaway activity and explosive metabolic requirements.

It is important to note that relatively small networks of neural elements, as connected as they are in the brain, are capable of generating these dynamical patterns. There is no need to require any external mechanisms to generate these attractor dynamics; they are likely to arise from the intrinsic activity of self-contained networks (Hertz, 1995; Wang & Blum, 1995). In particular, it is not necessary to assume between-stage feedback in order to produce desirable attractor dynamics, although between-stage feedback might well exist anyway (and be a further basis for attractor dynamics).

Therefore, existing knowledge of neural networks offers the possibility that they can behave in ways that are useful building blocks for higher order cognitive elements and operations.

FUTURE ISSUES AND SUMMARY

This review has moved from a gross behavioral and anatomic dissection of visual confrontation naming down to its possible roots in comparatively small neural networks. There were many topics relevant to theories of naming that could not be covered in this review, but that demand consideration in any complete theory of naming, such as, individual differences in function and anatomy; within-individual control of strategies and parameters of the naming process; priming and other forms of learning and memory; category-specific deficits (both within- and between-stage); the relationship of visual confrontation naming to other tasks, such as the naming of faces, word recall in spontaneous speech, and to reading; the neurodynamics of within-stage processing in naming and related cognitive functions; the effects of cerebral damage; and, ultimately, rehabilitation and drug effects on naming and name retrieval. It is hoped, though, that this review has better exposed the foundations on which theories of naming rest, and the points where contact with other fields and other tasks will be most informative.

ACKNOWLEDGMENTS

This work was supported in part by National Institute of Health (NIH) grants NS26553 and NS29973, by the McDonnell-Pew Program in Cognitive Neuroscience, by the Seaver Foundation, and by the Benjamin A. Miller Family Fund. I thank Harold Goodglass not only for his invitation, inspiration, and advice, but for his preternatural patience. I also thank Dana Boatman, Nathan Crone, John Hart, and Ola Selnes for helpful discussions, Gary Dell and Nadine Martin for permission to cite their unpublished work, and particularly Willem J. M. Levelt for insightful comments about testing models of naming, and for making available his position papers and lecture notes.

REFERENCES

Abbott, L. F., Rolls, E. T., & Tovee, M. J. (1996). Representational capacity of face coding in monkeys. *Cerebral Cortex, 6,* 498–505.

Arbib, M. A., Bischoff, A., Fagg, A. H., & Grafton, S. T. (1995). Synthetic PET: Analyzing large-scale properties of neural networks. *Human Brain Mapping, 2,* 225–233.

Berman, S., Friedman, D., Hamberger, M., & Snodgrass, J. G. (1989). Developmental picture norms: Relationships between name agreement, familiarity, and visual complexity for child and adult ratings of two sets of line drawings. *Behavior Research Methods, Instruments, and Computers, 21,* 371–382.

Berthier, M. L., Starkstein, S. E., Leiguarda, R., Ruiz, A., Mayberg, H. S., Wagner, H., Price, T. R., & Robinson, R. G. (1991). Transcortical aphasia: Importance of nonspeech dominant hemisphere in language repetition. *Brain, 114,* 1409–1427.

Besner, D. (in press). Basic processes in reading: Multiple routines in localist and connectionist models. In P. A. McMullen & R. M. Klein (Eds.), *Converging methods for understanding reading and dyslexia.* Cambridge, MA: MIT Press.

Binder, J. R., Swanson, S. J., Hammeke, T. A., Morris, G. L., Mueller, W. M., Fischer, M., Benbadis, S., Frost, J. A., Rao, S. M., & Haughton, V. M. (1996). Determination of language dominance using functional MRI: A comparison with the Wada test. *Neurology, 46,* 978–984.

Bookheimer, S. Y., Zeffiro, T. A., Blaxton, T., Gaillard, W., Theodore, W. (1995). Regional cerebral blood flow during object naming and word reading. *Human Brain Mapping, 3,* 93–106.

Borod, J. C., Goodglass, H., & Kaplan, E. (1980). Normative data on the Boston Diagnostic

Aphasia Examination, Parietal Lobe Battery, and the Boston Naming Test. *Journal of Clinical Neuropsychology, 2,* 209–216.

Breland, H. M. (1996). Word frequency and word difficulty: A comparison of counts in four corpora. *Psychological Science, 7,* 96–99.

Brennen, T., David, D., Fluchaire, I., & Pellat, J. (1996). Naming faces and objects without comprehension—A case study. *Cognitive Neuropsychology, 13,* 93–110.

Buchanan, L., & Besner, D. (1993). Reading aloud: Evidence for the use of a whole word non-semantic pathway. *Canadian Journal of Experimental Psychology, 47,* 133–152.

Bullock, T., McClune, M., Achimowicz, J., Iragui-Madoz, V., Duckrow, R., & Spencer, S. (1995). Temporal fluctuations in coherence of brain waves. *Proceedings of the National Academy of Sciences of the United States of America, 92,* 11568–11572.

Butler, A. B., & Hodos, W. (1996). *Comparative vertebrate neuroanatomy.* New York: Wiley-Liss.

Chiarello, C. (1991). Interpretation of word meanings by the cerebral hemispheres: One is not enough. In P. Schwanenflugel (Ed.), *The psychology of word meanings* (pp. 251–278). Hillsdale, NJ: Erlbaum.

Coltheart, M., Curtis, B., Atkins, P., & Haller, M. (1993). Models of reading aloud: Dual-route and parallel-distributed-processing approaches. *Psychological Review, 100,* 589–608.

Coney, J., & Abernethy, M. (1994). Picture–name priming in the cerebral hemispheres. *Brain and Language, 47,* 300–313.

Coslett, H. B., & Monsul, N. (1994). Reading with the right hemisphere: Evidence from transcranial magnetic stimulation. *Brain and Language, 46,* 198–211.

Crone, N. E., Hart, J., Boatman, D., Lesser, R. P., & Gordon, B. (1994). Regional cortical activation during language and related tasks identified by direct cortical electrical recording. *Brain and Language, 47,* 466–468.

Damasio, H., Grabowski, T. J., Tranel, D., Hichwa, R. D., & Damasio, A. R. (1996). A neural basis for lexical retrieval. *Nature, 380,* 499–505.

Dell, G. S. (1986). A spreading-activation theory of retrieval in sentence production. *Psychological Review, 3,* 283–321.

Dell, G. S. (1988). The retrieval of phonological forms in production: Tests of predictions from a connectionist model. *Journal of Memory and Language, 27,* 124–142.

Dell, G. S. (1990). Effects of frequency and vocabulary type on phonological speech errors. *Language and Cognitive Processes, 5,* 313–349.

Dell, G. S., & O'Seaghdha, P. G. (1991). Mediated and convergent lexical priming in language production: A comment on Levelt et al. (1991). *Psychological Review, 98,* 604–614.

Dell, G. S., & O'Seaghdha, P. G. (1992). Stages of lexical access in language production. *Cognition, 42,* 287–314.

Dell, G. S., Schwartz, M. F., Martin, N., Saffran, E. M., & Gagnon, D. A. (in press). Lexical access in aphasic and nonaphasic speakers. *Psychological Review.*

Desimone, R. (1991). Face-selective cells in the temporal cortex of monkeys. *Journal of Cognitive Neuroscience, 3,* 1–8.

Donders, F. C. (1868/1969). On the speed of mental processes. In W. G. Koster (Ed.), *Attention and Performance 2.* Amsterdam: North-Holland.

Dunn, L. M., & Dunn, L. M. (1981). *Peabody Picture Vocabulary Test—Revised.* Minnesota: AGS.

Feyereisen, P., Van Der Borght, F., & Seron, X. (1988). The operativity effect in naming: A reanalysis. *Neuropsychologia, 26,* 401–415.

Garrett, M. F. (1980). Levels of processing in sentence production. In B. Butterworth (Ed.), *Language production* (Vol. 1). New York: Academic Press.

Gernsbacher, M. A. (1984). Resolving 20 years of inconsistent interactions between lexical familiarity and orthography, concreteness, and polysemy. *Journal of Experimental Psychology: General, 2,* 256–281.

Glaser, W. R. (1992). Picture naming. *Cognition, 42,* 61–105.

Glosser, G., & Goodglass, H. (1991). Idiosyncratic word associations following right hemisphere damage. *Journal of Clinical and Experimental Neuropsychology, 13,* 703–710.

Goodglass, H. (1993). *Understanding aphasia.* San Diego: Academic Press.

Goodglass, H., & Baker, E. (1976). Semantic field, naming, and auditory comprehension in aphasia. *Brain and Language, 3,* 359–374.

Goodglass, H., Barton, M. I., & Kaplan, E. F. (1968). Sensory modality and object-naming in aphasia. *Journal of Speech and Hearing Research, 3,* 488–496.

Goodglass, H., Kaplan, E., & Weintraub, S. (1983). *Boston Naming Test.* Philadelphia: Lea and Febiger.

Goodglass, H., Wingfield, A., Bowles, N. L., Silberberg, M., Gallagher, R. E., Gleason, J. B., & Hyde, M. R. (1991, October). *The time course of semantic and phonological activation after presentation of an object picture.* Paper presented at the Academy of Aphasia Meeting, Rome, Italy.

Goodglass, H., Theurkauf, J. C., & Wingfield, A. (1984). Naming latencies as evidence for two nodes of lexical retrieval. *Applied Psycholinguistics, 5,* 135–146.

Gordon, B. (1982). Confrontation naming: Computational model and disconnection simulation. In M. A. Arbib, D. Caplan, & J. C. Marshall (Eds.), *Neural models of language processes* (pp. 511–530). New York: Academic Press.

Gordon, B. (1983). Lexical access and lexical decision: Mechanisms of frequency sensitivity. *Journal of Verbal Learning and Verbal Behavior, 22,* 24–44.

Gordon, B., Hart, J., Lesser, R., Schwerdt, P., Bare, M., Selnes, O., Fisher, R., & Uematsu, S. (1991). Visual confrontation naming mapped with direct cortical electrical stimulation [abstract]. *Neurology, 41 (suppl. 1),* 186–187.

Gordon, B., Hart, J., Lesser, R. P., & Selnes, O. A. (1994). Recovery and its implications for cognitive neuroscience. *Brain and Language, 47,* 521–524.

Hadar, U., Ticehurst, S., & Wade, J. P. (1991). Crossed anomic aphasia: Mild naming deficits following right brain damage in a dextral patient. *Cortex, 27,* 459–468.

Hart, J., Berndt, R. S., & Caramazza, A. (1985). Category-specific naming deficit following cerebral infarction. *Nature, 316,* 439–440.

Hart, J., Jr., & Gordon, B. (1990). Delineation of single-word comprehension deficits in aphasia, with anatomical correlation. *Annals of Neurology, 27,* 226–231.

Hart, J., Jr., & Gordon, B. (1992). Neural subsystems of object knowledge. *Nature, 359,* 60–64.

Hart, J., Crone, N. E., Lesser, R. P., Sieracki, J., Miglioretti, D. L., Hall, C., Sherman, D., & Gordon, B. (1997). Temporal dynamics of language comprehension. Unpublished manuscript.

Heilman, K. M., Tucker, D. M., & Valenstein, E. (1976). A case of mixed transcortical aphasia with intact naming. *Brain, 99,* 415–426.

Henderson, V. W. (1995). Naming and naming disorders. In H. S. Kirshner (Ed.), *Handbook of neurological speech and language disorders* (pp. 165–185). New York: Marcel Dekker.

Hendriks, H., & McQueen, J. (1996). *Max-Planck-Institut für Psycholinguistik Annual Report 16, 1995.* Nijmegen: Max-Planck-Institut für Psycholinguistik.

Hertz, J. (1995). Computing with attractors. In M. A. Arbib (Ed.), *The handbook of brain theory and neural networks* (pp. 230–234). Cambridge, MA: MIT.

Hillis, A., & Caramazza, A. (1995). The compositionality of lexical semantic representations: Clues from semantic errors in object naming. *Memory, 3/4,* 333–358.

Hinton, G. E., & Shallice, T. (1991). Lesioning an attractor network: Investigations of acquired dyslexia. *Psychological Review, 98,* 74–95.

Hirsch, K. W., & Ellis, A. W. (1994). Age of acquisition and lexical processing in aphasia: A case study. *Cognitive Neuropsychology, 11,* 458.

Hirsh, K. W., & Funnell, E. (1995). Those old, familiar things: Age of acquisition, familiarity and lexical access in progressive aphasia. *Journal of Neurolinguistics, 9,* 23–32.

Howard, D. (1995). Lexical anomia: Or the case of the missing lexical entries. *The Quarterly Journal of Experimental Psychology, 48A,* 999–1023.

Howes, D. (1966). A word count of spoken English. *Journal of Verbal Learning and Verbal Behavior, 5,* 572–606.

Humphreys, G., Riddoch, M., & Quinlan, P. (1988). Cascade processes in picture identification. *Cognitive Neuropsychology, 5,* 67–104.

Jacobs, R. A., & Jordan, M. I. (1992). Computational consequences of a bias toward short connections. *Journal of Cognitive Neuroscience, 4,* 323–336.

Jacobs, R. A., Jordan, M. I., & Barto, A. G. (1991). Task decomposition through competition in a modular connectionist architecture: The what and where vision tasks. *Cognitive Science, 15,* 219–250.

Jescheniak, J. D., & Levelt, W. J. M. (1994). Word frequency effects in speech production: Retrieval of syntactic information and of phonological form. *Journal of Experimental Psychology: Learning, Memory, and Cognition, 20,* 824–843.

Johnson, C. J., Paivio, A., & Clark, J. M. (1996). Cognitive components of picture naming. *Psychological Bulletin, 120,* 113–139.

Kaas, J. (1993). Evolution of multiple areas and modules within neocortex. *Perspectives on Developmental Biology, 1,* 101–107.

Kandel, E., & Abel, T. (1995). Neuropeptides, adenylyl cyclase, and memory storage. *Science, 268,* 825–826.

Kauffman, S. A. (1993). *The origins of order: Self-organization and selection in evolution.* New York: Oxford University Press.

Kay, J., & Ellis, A. (1987). A cognitive neuropsychological case study of anomia: Implications for psychological models of word retrieval. *Brain, 110,* 613–629.

Kempen, G., & Huijbers, P. (1983). The lexicalization process in sentence production and naming: Indirect election of words. *Cognition, 14,* 185–209.

Kirshner, H. S., Casey, P. F., Kelly, M. P., & Webb, W. G. (1987). Anomia in cerebral diseases. *Neuropsychologia, 25,* 701–705.

Kluender, K. R. (1994). Speech perception as a tractable problem in cognitive science. In M. A. Gernsbacher (Ed.), *Handbook of psycholinguistics* (pp. 173–217). San Diego: Academic Press.

Kosslyn, S. M., & Chabris, C. F. (1990). Naming pictures. *Journal of Visual Languages and Computing, 1,* 77–95.

Kremin, H. (1986). Spared naming without comprehension. *Journal of Neurolinguistics, 2,* 131–150.

Laine, M., & Martin, N. (1996). Lexical retrieval deficit in picture naming: Implications for word production models. *Brain and Language, 53,* 283–314.

Levelt, W. J. M. (1989). *Speaking: From intention to articulation.* Cambridge, MA: MIT Press.

Levelt, W. J. M., Schriefers, H., Vorberg, D., Meyer, A. S., Pechmann, T., & Havinga, J. (1991a). The time course of lexical access in speech production: A study of picture naming. *Psychological Review, 98,* 122–142.

Levelt, W. J. M., Vorberg, D., Pechmann, T., Schriefers, H., Meyer, A. S., & Havinga, J. (1991b). Normal and deviant lexical processing: Reply to Dell and O'Seaghdha (1991). *Psychological Review, 98,* 615–618.

Liu, Y. (1996). Queueing network modeling of elementary mental processes. *Psychological Review, 103,* 116–136.

Lueschow, A., Miller, E. K., & Desimone, R. (1994). Inferior temporal mechanisms for invariant object recognition. *Cerebral Cortex, 4,* 523–531.

Malow, B., Blaxton, T., Sato, S., Bookheimer, S., Kufta, C., Figlozzi, C., & Theodore, W. (1996). Cortical stimulation elicits regional distinctions in auditory and visual naming. *Epilepsia, 37*, 245–252.

Martin, A., Wiggs, C. L., Lalonde, F., & Mack, C. (1994). Word retrieval to letter and semantic cues: A double dissociation in normal subjects using interference tasks. *Neuropsychologia, 32*, 1487–1494.

Martin, A., Haxby, J., Lalonde, F., Wiggs, C., & Ungerleider, L. (1995). Discrete cortical regions associated with knowledge of color and knowledge of action. *Science, 270*, 102–105.

Martin, N., Gagnon, D. A., Schwartz, M. F., Dell, G. S., & Saffran, E. M. (1996). Phonological facilitation of semantic errors in normal and aphasic speakers. *Language and Cognitive Processes, 11*, 257–282.

Maunsell, J. (1995). The brain's visual world: Representation of visual targets in cerebral cortex. *Science, 270*, 764–769.

McClelland, J. L. (1979). On the time relations of mental processes: An examination of systems of processes in cascade. *Psychological Review, 86*, 287–330.

McClelland, J. L. (1993). The GRAIN model: A framework for modeling the dynamics of information processing. In D. E. Meyer & S. Kornblum (Eds.), *Attention and performance XIV.* Hillsdale, NJ: Erlbaum.

McCulloch, W. S., & Pitts, W. H. (1943). A logical calculus of the ideas immanent in nervous activity. *Bulletin of Mathematical Biophysics, 5*, 115–133.

McKeever, W. F., Seitz, K. S., Krutsch, A. J., & Van Eys, P. L. (1995). On language laterality in normal dextrals and sinistrals: Results from the bilateral object naming latency task. *Neuropsychologia, 33*, 1627–1635.

McNamara, T. P. (1992). Theories of priming: I. Associative distance and lag. *Journal of Experimental Psychology: Learning, Memory, & Cognition, 18*, 1173–1190.

Meyer, D. E., Irwin, D. E., Osman, A. M., & Kounios, J. (1988). The dynamics of cognition and action: Mental processes inferred from speed–accuracy composition. *Psychological Review, 95*, 183–237.

Miller, J. O. (1990). Discreteness and continuity in models of human information processing. *Acta Psychologica, 74*, 297–318.

Miller, J. O., & Hackley, S. A. (1992). Electrophysiologic evidence for temporal overlap among contingent mental processes. *JEP: General, 121*, 185–201.

Minsky, M., & Papert, S. (1969). *Perceptrons.* Cambridge, MA: MIT Press.

Morrison, C. M., & Ellis, A. W. (1995). Roles of word frequency and age of acquisition in word naming and lexical decision. *Journal of Experimental Psychology: Learning, Memory, and Cognition, 21*, 116–133.

Morrison, C. M., Ellis, A. W., & Quinlan, P. T. (1992). Age of acquisition, not word frequency, affects object naming, not object recognition. *Memory & Cognition, 20*, 705–714.

Morton, J. (1964). The effects of context on the visual duration threshold for words. *British Journal of Psychology, 55*, 165–180.

Morton, J., & Patterson, K. (1980). A new attempt at an interpretation, or, an attempt at a new interpretation. In M. Coltheart, K. Patterson, & J. C. Marshall (Eds.), *Deep dyslexia* (pp. 91–118). London: Routledge & Kegan Paul.

Nickels, L., & Howard, D. (1994). A frequent occurrence? Factors affecting the production of semantic errors in aphasic naming. *Cognitive Neuropsychology, 11*, 289–320.

Nickels, L., & Howard, D. (1995). Aphasic naming: What matters? *Neuropsychologia, 33*, 1281–1303.

Ohyama, M., Senda, M., Kitamura, S., Ishii, K., Mishina, M., & Terashi, A. (1996). Role of the nondominant hemisphere and undamaged area during word repetition in poststroke aphasics: A PET activation study. *Stroke, 27*, 897–903.

Ojemann, G. A. (1975). Language and the thalamus: Object naming and recall during and after thalamic stimulation. *Brain and Language, 2,* 101–120.

Ojemann, G. A., Creutzfeldt, O., Lettich, E., & Haglund, M. M. (1988). Neuronal activity in human lateral temporal cortex related to short-term verbal memory, naming and reading. *Brain, 111,* 1383–1403.

Ojemann, G., Ojemann, J., Lettich, E., & Berger, M. (1989). Cortical language localization in left, dominant hemisphere: An electrical stimulation mapping investigation in 117 patients. *Journal of Neurosurgery, 71,* 316–326.

Ojemann, G. A., Fried, I., & Lettich, E. (1989). Electrocortigraphic (ECoG) correlates of language. I. Desynchronization in temporal language cortex during object naming. *EEG Clinical Neurophysics, 73,* 453–463.

Patterson, K., Vargha-Khadem, F., & Polkey, C. E. (1989). Reading with one hemisphere. *Brain, 112,* 39–64.

Plaut, D. C., McClelland, J. L., Seidenberg, M. S., & Patterson, K. (1996). Understanding normal and impaired word reading: Computational principles in quasi-regular domains. *Psychological Review, 103,* 56–115.

Polk, T. A., & Farah, M. J. (1995). Brain localization for arbitrary stimulus categories: A simple account based on Hebbian learning. *Proceedings of the National Academy of Sciences (USA), 92,* 12370–12373.

Raff, R. A. (Ed.). (1996). *The shape of life.* Chicago: The University of Chicago Press.

Reggia, J. A., D'Autrechy, L., Sutton, G. G., & Weinrich, M. (1992). A competitive distribution theory of neocortical dynamics. *Neural Computation, 4,* 287–317.

Reis, A., Guerreiro, M., & Castro-Caldas, A. (1994). Influence of educational level of non-brain-damaged subjects on visual naming capacities. *Journal of Clinical and Experimental Neuropsychology, 16,* 939–942.

Riedl, R. D. (1978). *Order in living organisms: A systems analysis of evolution.* Chichester: John Wiley & Sons.

Ringo, J. L. (1991). Neuronal interconnection as a function of brain size. *Brain Behavior Evolution, 38,* 1–6.

Rochford, G., & Williams, M. (1962). Studies in the development and breakdown in the use of names, I: The relationship between nominal dysphasia and the acquisition of vocabulary in childhood. *Journal of Neurology, Neurosurgery, and Psychiatry, 25,* 222–233.

Roelofs, A. (1992). A spreading-activation theory of lemma retrieval in speaking. *Cognition, 42,* 107–142.

Rolls, E. T., & Tovee, M. J. (1995). The sparseness of the neural representation of stimuli in the primate temporal visual cortex. *Journal of Neurophysiology, 73,* 713–726.

Saffran, E. M., Schwartz, M. F., & Marin, O. S. M. (1976). Semantic mechanisms in paralexia. *Brain and Language, 3,* 255–265.

Salmelin, R., Hari, R., Lounasmaa, O. V., & Sams, M. (1994). Dynamics of brain activation during picture naming. *Nature, 368,* 463–465.

Sarter, M., Bernston, G., & Cacioppo, J. (1996). Brain imaging and cognitive neuroscience: Toward strong inference in attributing function to structure. *American Psychologist, 51,* 13–21.

Schriefers, H., Meyer, A. S., & Levelt, W. J. M. (1990). Exploring the time course of lexical access in language production: Picture-word interference studies. *Journal of Memory and Language, 29,* 86–102.

Seidenberg, M. S., & McClelland, J. L. (1989). A distributed, developmental model of word recognition and naming. *Psychological Review, 96,* 523–568.

Selnes, O. A., Risse, G. L., Rubens, A. B., & Levy, R. (1982). Transient aphasia with persistent apraxia. Uncommon sequela of massive left hemisphere stroke. *Archives of Neurology, 39,* 122–126.

Sergent, J. (1987). Information processing and laterality effects for object and face perception. In G. W. Humphreys & M. J. Riddoch (Eds.), *Visual object processing: A cognitive neuropsychological approach* (pp. 145–173). Hillsdale, NJ: Erlbaum.

Shepard, R. N. (1990). Neural nets for generalization and classification: Comment on Staddon and Reid. *Psychological Review, 97,* 579–580.

Shuren, J., Geldmacher, D., & Heilman, K. M. (1993). Nonoptic aphasia: Aphasia with preserved confrontation naming in Alzheimer's disease. *Neurology, 43,* 1900–1906.

Simon, H. A. (1962). The architecture of complexity. *Proceedings of the American Philosophical Society, 106,* 467–482.

Singer, W. (1993). Synchronization of cortical activity and its putative role in information processing and learning. *Annual Review of Physiology, 55,* 349–374.

Singer, W. (1995). Development and plasticity of cortical processing architectures. *Science, 270,* 758–764.

Smid, H. G. O. M., Mulder, G., Bocker, K. B. E., van Touw, A., & Brunia, C. H. M. (1996). A psychophysiological investigation of the selection and the use of partial stimulus information in response choice. *Journal of Experimental Psychology: Human Perception and Performance, 22,* 3–24.

Snodgrass, J. G., & McCullough, B. (1986). The role of visual similarity in picture categorization. *JEP: Learning, Memory, & Cognition, 12,* 147–154.

Snodgrass, J. G., & Vanderwart, M. A. (1980). A standardized set of 260 pictures: Norms for name agreement, familiarity and visual complexity. *Journal of Experimental Psychology: Human Learning and Memory, 6,* 174–215.

Starreveld, P. A., & La Heij, W. (1996). Time-course analysis of semantic and orthographic context effects in picture naming. *Journal of Experimental Psychology: Learning, Memory and Cognition, 22,* 896–918.

Stone, G. O., & Van Orden, G. C. (1989). Are words represented by nodes? *Memory & Cognition, 17,* 511–524.

Stone, G. O., & Van Orden, G. C. (1994). Building a resonance framework for word recognition using design and system principles. *Journal of Experimental Psychology: Human Perception and Performance, 20,* 1248–1268.

Swindale, N. V. (1990). Is the cerebral cortex modular? *Trends in Neurosciences, 13,* 487–492.

Tanaka, K. (1993). Neuronal mechanisms of object recognition. *Science, 262,* 685–688.

Terrace, H. S. (1985). In the beginning was the "name." *American Psychologist, 40,* 1011–1028.

Theios, J., & Amrhein, P. C. (1989). Theoretical analysis of the cognitive processing of lexical and pictorial stimuli: Reading, naming, and visual and conceptual comparisons. *Psychological Review, 96,* 5–24.

Thorpe, S., & Fize, D., & Marlot, C. (1996). Speed of processing in the human visual system. *Nature, 381,* 520–522.

Tippett, L. J., & Farah, M. J. (1994). A computational method of naming in Alzheimer's disease: Unitary or multiple impairments? *Neuropsychology, 8,* 3–13.

Triesman, A. M. (1960). Contextual cues in selective listening. *Quarterly Journal of Experimental Psychology, 12,* 242–248.

Tweedy, J. R., & Schulman, P. D. (1982). Toward a functional classification of naming impairments. *Brain and Language, 15,* 193–206.

Ullman, S. (1995). Sequence seeking and counter streams: A computational model for bidirectional information flow in the visual cortex. *Cerebral Cortex, 5,* 1–11.

Uttal, W. (1990). On some two-way barriers between models and mechanisms. *Perception & Psychophysics, 48,* 188–203.

Van Essen, D. C. (1979). Visual areas of the mammalian cerebral cortex. *Annual Review of Neuroscience, 2,* 227–263.

van Turennout, M., Hagoort, P., & Brown, C. M. (in press). Electrophysiological evidence on the time course of semantic and phonological processes in speech production. *JEP: Learning, Memory, and Cognition.*

von der Malsburg, C. (1995). Binding in models of perception and brain function. *Current Opinion in Neurobiology, 5,* 520–526.

Wang, X., & Blum, E. K. (1995). Dynamics and bifurcation of neural networks. In M. A. Arbib (Ed.), *The handbook of brain theory and neural networks* (pp. 339–343). Cambridge, MA: MIT.

Wang, X., Merzenich, M., Sameshima, K., & Jenkins, W. (1995). Remodelling of hand representation in adult cortex determined by timing of tactile stimulation. *Nature, 378,* 71–75.

Wang, G., Tanaka, K., & Tanifuji, M. (1996). Optical imaging of functional organization in the monkey inferotemporal cortex. *Science, 272,* 1662–1665.

Wheeldon, L. R., & Monsell, S. (1992). The locus of repetition priming of spoken word production. *Quarterly Journal of Experimental Psychology, 44A,* 723–762.

Wingfield, A., Goodglass, H., & Smith, K. L. (1990). Effects of word-onset cuing on picture naming in aphasia: A reconsideration. *Brain and Language, 39,* 373–390.

Zecevic, D. (1996). Multiple spike-initiation zones in single neurons revealed by voltage-sensitive dyes. *Nature, 381,* 322–325.

Chapter 3

On the Neurology of Naming

Daniel Tranel, Hanna Damasio, and Antonio R. Damasio

INTRODUCTION

Elsewhere we have proposed that the retrieval of word-form information, on the basis of which naming can occur, depends on the transient reactivation of the phonemic and morphologic structure of given words within the appropriate early sensory cortices (e.g., auditory, somatosensory, visual) and motor-related structures (Damasio, 1990; Damasio & Damasio, 1992). The traditional Wernicke's and Broca's areas are partly contained within such sensory and motor structures.

We have also proposed that the phonemic–morphologic reactivation pertaining to a given word is directed from a variety of regions located in higher-order association cortices, but largely outside the early sensory and motor sites alluded to above, and, consequently, largely outside the traditional language areas. Those regions operate as "third-party" neural mediators between, on the one hand, the regions that support conceptual knowledge (which are distributed over varied association cortices), and, on the other, the sensorimotor regions in which the phonemic–morphologic structure can be transiently reconstructed during the word-recall process, or instantiated during the perception of a word.

More specifically, we believe that, for most individuals, these regions are located in the left hemisphere, and are used as *lexical mediation units.* Once activated by the evocation of a given concept, these units, which do not contain phonemic–morphologic information in explicit form, promote the activation of the linguistic information necessary to reconstruct a word form, momentarily, in sensorimotor terms. These units also promote activation of syntactical information necessary for the proper placement of a word in the phrases and sentences being planned. In schematic form then, this is a tripartite model in which mediational structures interlink between conceptual and implementation structures, in two directions: from conceptual structure to the word-form reconstruction required for language acts, and in reverse, from the perception of a word towards its usual con-

ANOMIA: Neuroanatomical and Cognitive Correlates

ceptual structure correspondences. Our account is compatible with Levelt's (1989, 1992) proposal of lexical mediation units termed *lemmas,* based on cognitive experiments in normal subjects, which has also been applied to retrieval of words for actions (Roelofs, 1993).

The background for these hypotheses and for their investigation is the framework for synchronous retroactivation from convergence–divergence zones. The framework posits that retrieval of concepts depends on the reconstruction of images or actions pertaining to characteristics of entities. The images and movements themselves are reconstructed transiently in sectors of early sensory and motor structures, but the reconstruction is directed from separate system components in high-order association cortices, which contain dynamic regions that interlock feedforward and feedback projection neurons. Those regions, known as convergence–divergence zones, thus hold a preferential but probability-driven and modifiable dispositional capacity to signal directly or intermediately to sensory or motor regions, whenever the convergence–divergence site receives appropriate signals (Damasio, 1989a,b; Damasio & Damasio, 1994).

In the pages ahead, we review the new work from our laboratory that has led to these conclusions, beginning with studies pertaining to words for concrete entities, and then moving on to studies pertaining to words for actions. In order to place our work in context, we begin each section with a review of relevant studies from the literature. We then present results from new lesion studies, and finally, we summarize recent results obtained with functional imaging.

Some comments on terminology are in order at this point. When we use the designation *concrete entities,* we mean persons, places, animals, tools, and other items, that belong to varied conceptual classes, and that can be designated by an appropriately specific word (a proper or common noun). Concrete entities are mapped in the brain at different levels of *contextual complexity.* This permits us to classify a given entity along a dimension that ranges from *unique* (an entity belonging to a class with $N = 1$ and depending on a highly complex context for its definition), to varied *nonunique* levels (entities processed as belonging to classes with $N > 1$, having many members whose definition depends on less complex contexts). It is also important to distinguish between retrieval of conceptual knowledge about entities, on the one hand, which relates to the traditional term *recognition* (i.e., "knowing" what an item is), and retrieval of the word forms for entities, which refers to the traditional term *naming.* In our experiments, we required subjects to recognize and name unique entities (persons) at a *subordinate* level, and to recognize and name nonunique entities at a *basic object* level (cf. Rosch, Mervis, Gray, Johnson, & Boyes-Braem, 1976). The hypotheses were approached by studying patients with neurological lesions in various components of the large-scale neural system presumed to be related to the language lexicon. The demonstration of a defect in the retrieval of words relative to particular categories of entities was used as evidence for the relation between the putative system and the access to the lexicon for that particular conceptual category in the normal human brain.

RETRIEVAL OF WORDS FOR NONUNIQUE CONCRETE ENTITIES

Literature Review

For over a decade, investigators in several laboratories have noted intriguing dissociations in naming ability in patients with and without aphasia (Basso, Capitani, & Liacona, 1988; Damasio, 1990; Damasio, Damasio, & Van Hoesen, 1982; Franklin, Howard, & Patterson, 1995; Goodglass & Budin, 1988; Goodglass, Wingfield, Hyde, & Theukauf, 1986; Hart, Berndt, & Caramazza, 1985; Pietrini et al., 1988; Warrington & McCarthy, 1983; see also Small, Hart, Nguyen, & Gordon, 1995). These observations suggested that, following a lesion, not all lexical categories were equally compromised, thus opening the possibility that there were different neural systems required for the retrieval of words for different classes of concepts.

One frequent finding has been that brain-damaged patients demonstrate greater impairment in naming living (natural, animate) entities, as compared to artifactual (nonliving, human-made, inanimate) ones (Basso et al., 1988; Hart & Gordon, 1992; Hillis & Caramazza, 1991; Pietrini et al., 1988; Satori & Job, 1988; Silveri & Gainotti, 1988). In a few cases, the defect for living entities has been reported as being worse for one particular category—for instance, it has been found that either the category of animals (Hart & Gordon, 1992) or the category of fruits and vegetables (Farah & Wallace, 1992; Hart et al., 1985) was relatively more affected. In some cases, such patterns have been reported in connection with name comprehension: patients had disproportionate impairment in the comprehension of names of living entities compared to comprehension of names of artifacts (McCarthy & Warrington, 1988; Warrington & McCarthy, 1987). A few cases have been reported in which the dissociation occurred in the other direction; artifacts were disproportionately impaired, and living entities were relatively spared (Hillis & Caramazza, 1991; Sacchett & Humphreys, 1992; Warrington & McCarthy, 1983). Other patterns of defects have also been reported: disproportionate impairment of name comprehension for body parts (Goodglass & Budin, 1988), or relatively spared naming of body parts (Goodglass et al., 1986).

Another intriguing finding has been that patients may have their ability to process abstract words more preserved than their ability to process concrete words (Warrington, 1975, 1981; Warrington & Shallice, 1984). The reverse pattern has recently been reported; a patient whose defect in word retrieval was especially marked for abstract words and was less "anomic" for concrete words (Franklin et al., 1995).

Few of the studies cited above were driven by a hypothesis. Most of the studies consisted of single-case reports, in which some category-related dissociation was uncovered. Testing was often limited by incomplete sampling of the relevant categories. Another factor that hinders the interpretation of previous findings is the inconsistent usage of *naming,* on the one hand, and *recognition,* on the other. These capacities are quite separable, as can be easily noted by observing their frequent dissociation in brain-damaged patients. Nonetheless, some investigators have used the two concepts interchangeably (e.g., Riddoch & Humphreys, 1987; Sacchett & Humphreys, 1992), and it is thereby difficult to understand what was actually wrong with the patient. Finally, and of considerable importance as far as the neural basis of lexical retrieval is concerned, neuroanatomical data in these studies have been quite limited. In most cases, the neuroanatomical status of the patient was mentioned only in passing, and the lesions were not analyzed systematically. Subjects were often grouped together because they shared some aspect of their cognitive profiles, regardless of their lesion status, and inspection of such groups frequently reveals the subjects to have lesions in many different areas of the brain, even different hemispheres.

We reported a systematic study regarding the issue of category-related word-retrieval defects (Damasio, Damasio, Tranel, & Brandt, 1990; also see Damasio, 1990). The hint that there are separate neural systems involved in the naming performance for different lexical categories was supported. Moreover, in another study from our laboratory, we were able to obtain post mortem evidence for selective involvement of left temporal cortices in a case of progressive loss of word retrieval for concrete entities (Graff-Radford et al., 1990). We turn now to recent studies from our laboratory, which have confirmed and extended the initial findings.

New Lesion Study

We have added to previous findings on the breakdown of retrieval of words for nonunique entities by pursuing systematically their possible neural correlates (Damasio, 1992; Damasio, Brandt, Tranel, & Damasio, 1991; Damasio & Damasio, 1992; Damasio et al., 1990; Damasio & Tranel, 1990, 1993; Tranel, 1991).

In the studies summarized below, we tested the following hypotheses:

(1) Retrieval of words denoting concrete entities depends on neural structures that are anatomically distinct from the neural structures on which concept retrieval (recognition) for the same entities depends; and (2) Retrieval of words related to separate conceptual categories of concrete entities is mediated by distinct neural systems. We are not postulating the existence of "centers" that would hold a permanent memory (in this case, the memory of a word in the lexicon), but rather, the existence of dynamic systems containing dispositional knowledge on the basis of which the reconstruction of word images or articulatory patterns for those words can be directed and accomplished elsewhere, namely, in early sensory cortices and motor structures.

Subjects

Brain-damaged subjects. We conducted visual-recognition and naming experiments in 127 brain-damaged subjects with single, unilateral lesions, in order to address the hypotheses stated above. One hundred and nineteen were right-handed, five were left-handed, and three had mixed-handedness. The subjects were selected from the Division's Patient Registry so that as a group, they would permit us to sample the entire telencephalon. The subjects had lesions located in the left or right hemisphere, and in varied regions of the cerebral cortex, caused by either cerebrovascular disease, herpes simplex encephalitis, or temporal lobectomy. All had IQs in the average range or higher; had a high school education or higher; had been extensively characterized neuropsychologically and neuroanatomically; and had no difficulty with the attentive inspection of visual stimuli. None of the subjects had severe aphasia at the time of these experiments, although some had recovered from severe aphasia.

Control subjects. Normal control subjects ($n = 55$) were drawn from our visual-recognition and naming studies described in detail elsewhere (Damasio et al., 1990). They were matched to the brain-damaged subjects on age, education, and gender distribution. (Gender-related effects on visual recognition and naming are generally of fairly small magnitude [e.g., McKenna & Parry, 1994], with the most consistent finding being that women are better than men at naming fruits and vegetables, and men are better than women at naming animals. Hence, we used proportionate numbers of men and women in the brain-damaged and control groups, rather than analyzing the data separately by gender.)

Stimuli

The stimuli for the study were 300 nonunique entities, comprising 161 of the black-and-white line drawings from the Snodgrass and Vanderwart set (1980), and 139 additional black-and-white and color photographs. Five categories are represented in the stimulus set: animals ($n = 90$); fruits and vegetables ($n = 67$); tools and utensils ($n = 104$); vehicles ($n = 23$); and musical instruments ($n = 16$).

Procedure

The entities were depicted on slides and shown in random order one-by-one on a Caramate 4000 slide projector, in free field. For each, the subject was asked to tell the experimenter what the entity is ("What is this?"). If the subject produced a vague or superordinate-level response (e.g., "some kind of animal"), the subject was prompted to "be more specific; tell me exactly what you think that thing is." Time limits were not imposed. All responses were audiotaped.

Neuropsychological data quantification

For each stimulus, the response of each brain-damaged patient was scored as correct if it matched one of the responses accepted as correct from the normal controls. For each experiment, we first determined which stimuli the patient *recognized* (see Damasio et al., 1990). In brief, a recognition response was scored as correct if either of two conditions was met: (1) the stimulus was named correctly (we accept this as unequivocal evidence of correct recognition; it should be noted that we have never found a subject who would produce a correct name, and not recognize the stimulus that was named); or (2) the subject provided a specific description of the entity (e.g., "That's an animal that can store water in the hump on its back, lives in the desert, can go a long time without water, and can be ridden."). The number of stimuli the subject recognized correctly, divided by the total number of items in the category and multiplied by 100, constituted the *recognition score.*

The *naming score* was calculated by summing the number of correct naming responses using only those stimuli for which the subject had produced a correct *recognition* response. If a subject did not recognize a particular stimulus, that stimulus was not included in the naming score calculation. In this approach to data quantification, subjects are not penalized for failing to name stimuli that they also do not recognize. Scores were multiplied by 100 to produce final figures in terms of percent correct.

Neuroanatomical methods

The neuroanatomical analysis was based on magnetic resonance imaging (MRI) data, or in those subjects in whom an MRI could not be obtained, on computerized axial tomography (CT) data, obtained in the chronic epoch (at least 3 months postonset of lesion).

In most subjects, MRI scans were obtained with an SPg sequence of thin (1.5 mm) and contiguous T_1-weighted coronal cuts. The resulting 124 slices were processed in Brainvox (H. Damasio & Frank, 1992) and a three-dimensional (3-D) reconstruction was obtained for each individual subject. In a few cases only a standard MRI sequence was available with both axial and coronal T_1-weighted, 5-mm thick slices. The anatomical description of the lesion and of its placement relative to neuroanatomical landmarks was performed at the workstation screen taking advantage of Brainvox routines (e.g., rotation of 3-D volume; 2-D to 3-D transfer; reslicing capability; image enlargement; image lighting). The results of the analysis were stored in relation to the standard brain segmentation used in our laboratory (Damasio & Damasio, 1989).

In a subsequent step, all lesions in this set were transposed and warped into a normal 3-D brain, so as to permit the determination of the maximal overlap of lesions relative to subjects grouped by neuropsychological defect. This technique, known as MAP-3, is carried out as follows: The normal 3-D brain is resliced so as to match the slices of the MRI/CT of the subject and create a correspondence between each of the subject's MRI/CT slices and the normal resliced brain. The contour of the lesion on each slice is then transposed onto the matched slices of the normal brain, warping it in relation to the available anatomical landmarks. The summation of these contours defines an "object" that represents, for each subject, the lesion in three dimensions placed in the normal reference brain. The final step consists of the detection of the intersection of the several "objects" in the reference brain, and the analysis of their placement in relation to the anatomical detail of the reference brain. If the analysis is only concerned with the surface overlap of the lesions, we have designated it as MAP-2. In this instance, each view of the 3-D brain showing the lesion is matched, for each subject, with the corresponding view of the normal brain. The contour of the lesion is transposed from the subject's brain onto the normal brain taking into account its relation to sulcal and gyral landmarks. (MAP-2 can also be used in cases for which only a 2-D MRI or CT scan exists.)

Results concerning hypothesis #1. Retrieval of words versus retrieval of concepts for the same concrete entities.

Neuropsychological findings

Table 1 presents neuropsychological results pertinent to the question of whether retrieval of conceptual knowledge and retrieval of words can be dissociated. So far, we have uncovered three different patterns of recognition and naming profiles, with respect to nonunique concrete entities: (1) Group 1: defective naming accompanied by normal recognition; (2) Group 2: defective naming accompanied by defective recognition (subjects have defective recognition and in addition, have defective naming of items they can recognize); (3) Group 3: normal naming but defective recognition (as defined for Group 2) (i.e., subjects who could name correctly entities that

Table 1 Retrieval of Words and Concepts[a]

	Category					
	Animals		Fruits/Vegetables		Tools/Utensils	
	N	R	N	R	N	R
Group 1	**75.0**	90.9	**85.2**	91.7	**84.5**	95.7
(n = 10)	(15.0)	(3.7)	(16.4)	(3.4)	(8.9)	(2.8)
Group 2	**80.3**	80.6	**82.1**	84.9	**73.5**	83.0
(n = 12)	(20.6)	(15.3)	(17.5)	(12.5)	(21.9)	(9.7)
Group 3	94.7	**77.5**	94.6	**72.8**	95.7	94.8
(n = 21)	(3.2)	(8.0)	(4.9)	(19.7)	(3.3)	(3.4)
Normal controls	95.7	91.9	94.3	92.6	98.2	96.2
(n = 55)	(3.1)	(2.8)	(3.7)	(3.9)	(1.9)	(3.3)

[a] Means (standard deviations); N = percent correct naming; R = percent correct recognition (boldfaced scores are defective).

they recognized, even though overall recognition was defective in one or more categories). Data from the normal controls are also presented.

Neuroanatomical findings

Abnormal retrieval of words and normal retrieval of concepts (Table 1—Group 1).   The 10 subjects in this group had lesions in the left temporal lobe, mostly overlapping in the middle and lateral inferotemporal (IT) region, and in the lateral aspect of the temporo-parietal region.

Abnormal retrieval of words and concepts (Table 1—Group 2).   The 12 subjects in this group showed lesions mainly in the left lateral temporo-occipital region, mostly overlapping at the temporo-occipito-parietal junction, at the posterior end of the superior temporal sulcus.

Normal retrieval of words and abnormal retrieval of concepts (Table 1—Group 3).   Of the 21 subjects in this group, 11 had lesions in the right hemisphere, and 10 in the left. The right unilateral lesions were concentrated in the inferior and mesial aspects of the occipital lobe. The main overlap of lesions was in the infracalcarine region (lingual gyrus), tapering antero-inferiorly in the posterior IT region. In the left hemisphere, there was an overlap of subjects in the anterior sector of the fusiform gyrus.

Results for Hypothesis #2.   Retrieval of words for concrete entities from separate conceptual categories.

Neuropsychological findings

In Table 2, the results are organized according to the category of naming impairment (collapsed across whether or not there was an accompanying recognition defect). In Group 1, subjects had an animal naming defect, irrespective of other naming impairments; in Group 2 (a subset of Group 1), the defect was *restricted* to the animal category. In Group 3, subjects had a tool and utensil naming defect, irrespective of other naming impairments; in Group 4 (a subset of Group 3), the defect was *restricted* to the tool and utensil category.

We conducted statistical comparisons, to confirm the reliability of the findings. Using *t*-tests, we compared the naming scores of defective groups to the naming scores of brain-damaged subjects who were not defective. The results supported the conclusion that there were significant differences between groups. For Group 1, the animal-naming score differed significantly from the score of nondefective subjects ($t(125) = -4.78, p < .001$). For Group 3, the tool–utensil naming score differed significantly from the score of nondefective subjects ($t(125) = -4.56, p < .001$).

Neuroanatomical findings

Abnormal retrieval of words for animals (Table 2—Group 1). A defect in retrieval of words for animals was observed in 16 subjects. In all but one, the lesions occurred in the left IT region. The maximal overlap was seen in the midanterior sector of the lateral and inferior aspect of IT. The overlap then tapered towards the anterior sector of IT and the temporal pole. With the exception of one subject, whose lesion was in the mesial left occipital region (in both supra- and infracalcarine regions), no lesions outside the IT region were associated with the defect.

Abnormal retrieval of words for tools and utensils (Table 2—Group 3). Abnormal retrieval of words for tools and utensils was associated with damage in the left lateral temporal and occipital region and, to a lesser extent,

Table 2 Word Retrieval Defects for Animals and/or Tools/Utensils[a]

| | Category | | |
Nature of defect	Animals	Fruits/Vegetables	Tools/Utensils
Group 1: Animal	**66.6**	70.6	74.9
($n = 16$)	(22.5)	(28.8)	(26.8)
Group 2: Animal only	**76.9**	90.3	92.3
($n = 7$)	(8.7)	(4.6)	(3.1)
Group 3: Tool/utensil	**74.1**	73.7	**68.2**
($n = 16$)	(26.8)	(29.2)	(23.5)
Group 4: Tool/utensil only	93.9	93.2	**76.9**
($n = 7$)	(3.2)	(7.7)	(7.3)

[a] Means (standard deviations); boldfaced scores are defective.

in the left parietal region. (One subject had a lesion in the right temporal pole.) The maximal lesion overlap occurred in the left temporo-occipito-parietal junction. It is interesting to note that this is largely the same place as the lesion overlap we have found for abnormal retrieval of *concepts* for tools and utensils (see Tranel, Damasio, Damasio, & Brandt, 1995). In seven subjects, the word-retrieval defect was restricted to tools and utensils, and did not affect animals (Table 2—Group 4).

Functional Imaging (PET) Study

We recently completed a functional imaging study regarding retrieval of words for concrete unique and nonunique entities (H. Damasio et al., 1996). We studied nine normal right-handed young adults, ranging in age from 22 to 49, all of whom were native English speakers. There were seven women and two men. The subjects engaged in three tasks: (1) naming unique persons from their faces; (2) naming animals; and (3) naming tools and utensils (results for the naming of unique familiar faces are reported in the section on Unique Concrete Entities below). In a control task, subjects were asked to decide and report whether unfamiliar faces were presented right-side up or upside down.

The naming tasks were performed during a positron emission tomography (PET) scanning session. Subjects performed each task twice, in random order. Task performance began 5 sec after injection of $[^{15}O]H_2O$ into the antecubital vein and continued until 65 sec after injection. Oral responses were recorded, and performance measures (accuracy, latency) were obtained. For each task, the stimuli were presented at set rates that pilot studies had shown to yield similar high but nonperfect performance accuracies (i.e., rates at which subjects could perform well, but not at ceiling). Specifically, the familiar faces were presented one every 2.5 sec, the tools and utensils were presented one every 1.8 sec, and the animals were presented one every 1.5 sec. These rates produced performance levels of about 90% correct for each task type. In a separate session, MRIs of each subject's brain were obtained and reconstructed in three dimensions using Brainvox (H. Damasio & Frank, 1992).

MRI and PET data were coregistered a priori using PET-Brainvox (Grabowski et al., 1995). This fit was corrected post hoc with Automated Image Registration (AIR; Woods, Mazziotta, & Cherry, 1993). PET data were subjected to Talairach transformation (Talairach & Szikla, 1967), based on analysis of the coregistered 3-D MRI data set. The data were analyzed with a pixelwise two-way analysis of covariance (ANCOVA) (estimated coefficients for global flow serving as the covariate), in which we compared adjusted mean activity in each of the three naming conditions to the control task (Friston, Frith, Liddle, & Frackowiak, 1991). Regions of statistically significant changes in normalized regional cerebral blood flow (rCBF) for each of the three naming tasks were searched for in the temporal polar (TP) area and in the IT cortices identified in the 3-D reconstructed MRI scans of each subject.

When subjects named animals and tools and utensils, there were significant increases in rCBF in distinct loci in the posterior IT region in the left hemisphere. For the tools and utensils, the principal location of activation was in the posterolateral aspect of left IT, in the middle and inferior temporal gyri. Naming of animals produced activation that was mesial and anterior to that produced by naming tools and utensils, in the inferior and fourth temporal gyri. Most importantly, the areas of activation produced by naming animals and tools and utensils, were not just separate, but also corresponded to the areas identified by the lesion studies as being crucial for these capacities. Our results are also consistent with another recent PET study concerned with similar issues (Martin, Haxby, Lalonde, Wiggs, & Ungerleider, 1995).

Conclusions

In the data described in the previous sections, we found support for the following conclusions:

1. The neural systems required to retrieve conceptual knowledge for nonunique entities and those required to retrieve the words for those entities, are separate, at least in part. This seems to be true for the animal category, but may not be true for the tool and utensil category.

2. The neural systems required to retrieve words for nonunique entities seem to be based nearly exclusively in *left* hemisphere regions. We found 1 subject (out of 127) in whom defective word retrieval occurred with a lesion in the right hemisphere.

3. Abnormal retrieval of words for animals was seen in subjects with lesions that clustered in IT. The maximal overlap occurred in lateral and inferior IT regions, in the anterior sector of the middle and inferior temporal gyri. TP was not included in the overlap. The subcortical overlap was subjacent to the cortical damage and extended posteriorly, lateral to the temporal horn and the lower segment of the trigone.

4. Abnormal retrieval of words for tools was seen in subjects whose lesions involved posterior and lateral temporal cortices and the supramarginal gyrus. The maximal overlap occurred at the back end of the middle temporal gyrus and the anteroinferior sector of the supramarginal gyrus. The subcortical overlap was distinct from the one seen in (3), by being both posterior and superior to it.

5. The two-system segregation effects noted above—words versus concepts and words for animals versus words for tools and utensils—seem to obey consistent principles. Specifically, relative to the stream architecture of cortical projection neurons in the occipitotemporal (OT) region: (a) word-retrieval defects for animals depend on damage to systems located anteriorly to those whose damage produces tool and utensil word-retrieval defects; (b) in no instance was there a defect for word retrieval for animals caused by a lesion posterior to a lesion causing word-retrieval defects for tools and utensils.

We have also found category-related neuropsychological and neuroanatomical dissociations regarding the retrieval of *conceptual knowledge* for concrete entities, which parallel to some extent the dissociations described above for word-form retrieval (Damasio, 1990; Damasio et al., 1990; Tranel, Damasio, & Damasio, 1995). Specifically, in a large-scale study of subjects with lesions throughout various sectors of the telencephalon, we found that damage centered in the right mesial and inferior OT region, or centered in the left mesial occipital region, produced impairments in the retrieval of conceptual knowledge (recognition) for animals. Damage centered in the left posterior temporo-occipital and parieto-occipital regions produced impairments in the retrieval of conceptual knowledge for tools and utensils. These findings indicate that, as in the case of word forms, the recording and retrieval of conceptual knowledge of different domains depend on partially segregated neural systems.

The results from our word-retrieval studies suggest that the sites we are identifying by means of lesion overlap and by functional imaging are critical to the reconstruction of explicit sensory or motor patterns, on the basis of which word forms are made available to consciousness. The sites we are identifying do not contain permanent and explicit representations of words, but rather, dispositions on the basis of which such explicit representations can be transiently activated in early sensory cortices or motor structures. Moreover, we do not see the sites we have identified as "centers" for such dispositions, but rather as the high-order components of a network that includes other dispositional sites and whose operation leads to the above-mentioned transient patterns of reconstructions (for theoretical background, see Damasio, 1989a,b; Damasio & Damasio, 1992, 1994).

RETRIEVAL OF WORDS FOR UNIQUE CONCRETE ENTITIES

Literature Review

In addition to differences in word-retrieval performance for nonunique entities of diverse categories, we and others have made the clinical observation that retrieval of words for unique, concrete entities (e.g., persons, places) may be disproportionately compromised relative to the retrieval of words for nonunique entities, and in some instances, may constitute the only word-retrieval defect (Carney & Temple, 1993; Cipolotti, McNeil, & Warrington, 1993; Cohen, Bolgert, Timsit, & Cherman, 1994; Damasio, 1990; Flude, Ellis, & Kay, 1989; Hittmair-Delazer, Denes, Semenza, & Mantovani, 1994; Lucchelli & De Renzi, 1992; McKenna & Warrington, 1978, 1980; McNeil, Cipolotti, & Warrington, 1994; Semenza & Sgaramella, 1993; Semenza & Zettin, 1988, 1989; Shallice & Kartsounis, 1993). As in the case of retrieval of words for nonunique entities, however, most of these studies were conducted with little or no regard for neuroanatomical factors.

As far as understanding the neural basis of retrieval of words for unique entities, the disregard for anatomical considerations has produced a rather confusing array of findings. In fact, Hittmair-Delazer et al. (1994) concluded after reviewing this literature that it was not possible to provide a neuroanatomical basis for "proper name" retrieval. The authors provided a table showing the various lesion locations for the cases of proper anomia that have been reported to date, and concluded that the "relative pureness of all reported cases" and the observation of "quite distinctly located anatomical lesions" made it unlikely that an anatomical explanation would suffice. However, we believe this interpretation is a consequence of either insufficient precision of neuroanatomical analysis or inadequate assessment of naming compromise, or perhaps both. In fact, we have new findings obtained with careful neuroanatomical and neuropsychological investigations in a large series of patients, which point strongly and consistently to a specific neural correlate for retrieval of words for unique entities (Damasio et al., 1990; Graff-Radford et al., 1990). The findings also suggest that access to words for unique entities depends on neural systems distinct from those that support access to words for nonunique entities (Damasio & Damasio, 1992; Damasio et al., 1991, 1995). The findings are summarized below.

New Lesion Study

The primary hypothesis we addressed in these experiments is that access to words for unique entities depends on neural systems distinct from the neural systems that support access to words for nonunique entities.

Subjects

Brain-damaged subjects. We addressed the hypothesis by studying subjects with damage to varied sectors of the left or right hemisphere, as described in the previous section (*N* = 127).

Control subjects. For the face-naming task described below, 60 normal control subjects were studied. There were 10 men and 10 women in each of three age brackets: (1) age 20–39; (2) age 40–59; and (3) age 60 and over.

Method

Stimuli from two tests were utilized to measure naming of famous faces: 77 items from the Iowa Famous Faces Test (Tranel, Damasio, & Damasio, 1995), and 56 items modified from the Boston Famous Faces Test (Albert, Butters, & Levin, 1979). The stimuli are black-and-white slides depicting famous actors, politicians, and sports figures. The slides were shown to the subject one at a time, and for each, the subject was asked to (1) indicate whether the face is familiar, and if so, to (2) indicate who the person is, and (3) supply the person's name. The percent correct score for the *naming* part of each test was the dependent measure.

Data quantification

For each test, the score of each brain-damaged subject was compared to the mean from the relevant control group. Scores were classified as "defective" if they were 2 or more standard deviations below the control mean.

Neuropsychological results

The principal results are presented in Table 3. The focus here is on retrieval of proper nouns, but we also present data from the common noun-retrieval tasks, to facilitate comparisons. Three sets of results are presented: Group 1: Subjects with defective retrieval of words for unique entities (regardless of whether nonunique word retrieval was normal or not). Group 2: Subjects with defective retrieval of words for unique entities, in whom retrieval of words for nonunique entities was normal. Group 3: Subjects with defective retrieval of words for nonunique entities, in whom retrieval of words for unique entities was normal. The data are presented in

Table 3　Defective Retrieval of Words for Unique Entities[a]

Group	Faces	Animals	Fruits/ Vegetables	Tools/ Utensils
Group 1 (n = 13)	13/13	6/13	6/13	4/13
Group 2 (n = 7)	7/7	0/7	0/7	0/7
Group 3 (n = 17)	0/17	10/17	10/17	12/17

[a]Number of defective subjects/number of subjects in group.

terms of the number of subjects in each group who expressed a *defect* in the relevant domain.

It is important to note that *none* of the subjects in any of these groups had a defect in the retrieval of conceptual knowledge for unique entities (i.e., their recognition of familiar persons was normal). In short, the defect (if present) was restricted to the retrieval of *names* for familiar persons; there was no impairment of recognition of identity.

Neuroanatomical results

The techniques for analysis were the same as those reported for the neuroanatomical results in Study 1 above regarding retrieval of words for nonunique entities. In the seven subjects with a *pure* defect in retrieval of words for unique entities (Group 2), damage was centered in the left TP region, in both inferomesial and lateral aspects. When all subjects with a deficit in retrieval of proper nouns were considered (i.e., Group 1, with six additional subjects who had defects in retrieval of words for animals and in some cases, for animals and tools and utensils), the site of maximal lesion overlap remained in the antero-lateral and inferior sectors of the left temporal pole. For Group 3, the lesions clustered in the left posterior OT region.

None of the other subjects (outside of Groups 1–3) showed a deficit in retrieval of words for unique entities. Of particular importance was the finding that damage to the *right* anterior temporal region did not produce the defect. Of seven subjects with lesions to this sector, none had an impairment in retrieval of words for unique entities.

Functional Imaging (PET) Study

The methods for the functional imaging study of proper naming were described in the previous section (H. Damasio et al., 1996), and we report here the relevant results. When the nine normal subjects named familiar faces, there was an increase in normalized rCBF in the left TP area, but not in left IT. The left TP activation corresponds to the same region identified in the lesion study as being important for the retrieval of words for unique entities. There was also activation in the right TP area, which we interpret to reflect the *recognition* of the unique faces that would inevitably accompany the naming of the faces (the control task did not require recognition at unique level). Again, this finding closely parallels results from lesion studies, as noted above. The TP activation associated with naming unique entities was distinct from the activation in IT associated with naming of nonunique entities. This finding, together with the lesion results described above, provides convergent evidence for the existence of distinct anatomical systems supporting retrieval of words for unique versus nonunique entities.

Conclusions

1. Lesions to the left TP region, both circumscribed or inclusive of the IT sector, were associated with defects in retrieval of words for unique concrete entities.

2. None of the subjects in (1) had a defect in retrieving the *concepts* of the persons for whom they could not recover the names (i.e., recognition of identity was normal).

3. Lesions restricted to the left IT region, which did not affect the temporal pole, did not cause defects in retrieval of words for unique entities, although an impairment in word retrieval for nonunique entities was ubiquitous in this group.

4. Right-sided TP lesions did not cause word-retrieval defects for unique entities.

5. The findings continue to obey the principles described in the conclusions of Study 1 regarding word retrieval for nonunique entities. Word retrieval for unique concrete entities depends on a system located anteriorly to the systems on which word retrieval for nonunique concrete entities depends. In no instance was there a violation of this principle, in no case was a naming defect for unique entities related to a lesion located posteriorly to a lesion that caused a defect in naming of nonunique entities (e.g., animals). Moreover, in no instance was there a combination of a defect in naming unique entities with a defect in naming nonunique entities of the tool and utensil variety, without a defect in naming animals as well. In no instance did the lesion placement we have identified (i.e., lesions related to deficits in the retrieval of words for unique entities, nonunique animals, and nonunique tools and utensils) fail to be respected.

6. The results from our lesion and PET studies are quite compatible. The activation sites for the naming of unique persons, and nonunique animals and tools and utensils, corresponded closely to those identified in the lesion studies (Tranel et al., 1995; Damasio et al., 1995; H. Damasio et al., 1996).

RETRIEVAL OF WORDS FOR ACTIONS: PRELIMINARY EVIDENCE

Literature Review

Our studies on word retrieval have now been extended to the class of *actions*. In a preliminary study (Damasio & Tranel, 1993), we reported a subject with a well-defined lesion in the left frontal operculum, involving both prefrontal and premotor cortex and underlying white matter. His pattern of performance showed defective retrieval of words for actions and normal retrieval of words for nonunique entities. This subject was contrasted with two other subjects who had left IT lesions. These two subjects had defective retrieval of words for nonunique entities but normal retrieval of words for actions. Regarding the first subject and the other two, the lesions were nonoverlapping and the neuropsychological performances were quite distinct. The result was a double dissociation relative to both naming performance and site of lesion.

This result constituted the first suggestion that the systems required for naming of entities and naming of actions are, at least in part, segregated in the human brain, even if they normally operate in coordinated fashion. Our result came on a background of observations that patients can have disproportionate impairment of the ability to retrieve "nouns" or "verbs" (Caramazza & Hillis, 1991; Hillis & Caramazza, 1995; McCarthy & Warrington, 1985; Miceli, Silveri, Noncentini, & Caramazza, 1988; Miceli, Silveri, Villa, & Caramazza, 1984; Zingeser & Berndt, 1988, 1990). None of these reports, however, aimed at presenting neuroanatomical data or hinted at a neural system segregation. Two recent reports (Daniele, Giustolisi, Silveri, Colosimo, & Gainotti, 1994; Miozzo, Soardi, & Cappa, 1994) support the Damasio and Tranel (1993) study in three and one patients, respectively, although both studies provide only a hint of neuroanatomical data.

Another recent study compared syntactic constructions between agrammatic (usually Broca's) aphasics and conduction aphasics (Goodglass, Christiansen, & Gallagher, 1994). Consistent with previous studies of this type (e.g., Marin, Saffran, & Schwartz, 1976), the authors found that agrammatic aphasics had a preponderance of nouns over verbs in running speech and in single-constituent utterances. This finding is consistent with our work (e.g., Damasio & Tranel, 1993), and with other studies reporting noun–verb discrepancies in word repetition (Katz & Goodglass, 1990) and written word retrieval (Baxter & Warrington, 1985; Caramazza & Hillis, 1991).

We have now replicated and extended our initial finding in several additional cases, and these new findings are summarized below.

New Lesion Study

Subjects

We studied subjects with damage to the left premotor and prefrontal region. Subjects with lesions to the left temporal, parietal, or occipital cortices have also been studied. Brain-damaged controls were subjects with right-hemisphere lesions. In all, we have studied 83 brain-damaged subjects and 10 normal older controls. The brain-damaged subjects were drawn from our Patient Registry, and they conformed to the same inclusion criteria specified for the study on nonunique entities described above (see the first New Lesion Study section). Of the brain-damaged subjects, 14 with lesions in "target" regions have undergone detailed neuroanatomical analysis, using the template method described in Damasio and Damasio (1989).

Experimental task

The task for measuring retrieval of words for actions was the Action Recognition and Naming Task recently developed and standardized in our laboratory (Fiez & Tranel, 1994; 1997). In brief, the test comprises 280 color photographs of various actions. The items elicit responses that vary along several dimensions, including: (1) the inflection of the elicited response (gerundial forms [e.g., *eating*] vs. past-tense forms [e.g., *ate*]), (2) the frequency of the elicited verb per million words (Francis & Kucera, 1982), (3) the type of agent performing each action (person, animal, or object), and (4) compatability with different argument structures; some elicited responses can only be produced in a well-formed sentence as an intransitive verb (one place predicate: e.g., "John RAN/IS RUNNING"), some only as a transitive verb (two-place predicate: e.g., "John HIT/IS HITTING the ball"; or three-place predicate: e.g., "John GAVE Mary a book), and some can be produced in either type of sentence. The items also represent a diverse range of conceptual categories (e.g., verbs of perception, motion, etc.). In the test, 75% of the stimuli are single pictures depicting an ongoing action, for which subjects are instructed to produce a single word that best describes what the person, animal, or object is doing (e.g., "walking"). The remaining 25% of the stimuli are picture pairs depicting some change in an object, and subjects are asked to produce a single word that best describes what was done to the object, or what the person or object did (e.g., "chopped").

Scoring and Data Quantification

The Action Recognition and Naming Test was standardized in a series of experiments conducted in normal subjects (Fiez & Tranel, 1997). This standardization yielded, for each item on the test, a naming response (or in a few instances, two or three responses), which was considered correct. To quantify the performances of the brain-damaged subjects, we compared their responses to those of the standardization sample, and scored as correct those responses that matched those produced by the normal controls.

Table 4   Summary of Action Naming Data[a]

Lesion group	Total N	Naming (and recognition)	Naming only
Left frontal	19	10	0
Left occipitotemporal	21	8	2
Right occipitotemporal	9	1	1
Other	34	2	1
TOTAL:	83	21	4

[a] Number of subjects defective.

A percent correct score was then calculated for each brain-damaged subject, and the score was classified as *defective* if it was two or more standard deviations below the mean of the control group.

A hierarchical approach was used to classify the brain-damaged subjects into four different groups. First, we identified a group of 19 subjects whose lesions included (but were not necessarily limited to) damage to any of the following left frontal areas: left frontal operculum, premotor region, Rolandic region, basal ganglia. From the remaining 64 subjects, we next identified a group of 21 subjects whose lesions included (but were not necessarily limited to) damage to any of the following left OT areas: infracalcarine cortex, supracalcarine cortex, the mesial temporo-occipital junction, the posterior portion of the middle temporal gyrus, the posterior portion of the inferior temporal gyrus. From the remaining 43 subjects, we next identified a group of 9 subjects whose lesions included (but were not necessarily limited to) damage to the mesial OT areas in the right hemisphere. The remaining 34 subjects (whose lesions did not extend into any of the areas listed above) were not classified into specific neuroanatomical subgroups for the analysis presented in Table 4.

Results

We found a number of subjects who demonstrated naming defects for actions. In most cases, the naming impairment was accompanied by a defect in the retrieval of conceptual knowledge for actions (the "Naming and Recognition" group), although we focus here primarily on the naming results. A summary of the findings, using the lesion analysis approach described above, is presented in Table 4.

Two preliminary conclusions can be drawn from these findings. First, the initial result regarding the association of action word-retrieval failure with damage to the left frontal region has been replicated in a number of additional cases (Damasio & Tranel, 1993). Second, there is a strong association between recognition and naming performances. For example, in the brain-damaged subjects with defective naming, we found that there is a significant correlation between naming performance and the average conceptual retrieval (recognition) performance ($r = .55$, $p < .005$). In other words, recognition and naming tend to go together quite strongly with regard to actions, which contrasts with the frequent dissociations we have obtained between these two capacities with regard to concrete entities, especially persons and animals.

To summarize, we found 10 subjects with a pattern of word retrieval that replicated the initial observation reported by Damasio and Tranel (1993) (i.e., abnormal retrieval of words for actions and normal retrieval of words for concrete nonunique entities). Neuroanatomical analysis in these subjects revealed that the lesions clustered in the inferior motor and premotor regions, with the maximal lesion overlap occurring in the inferior frontal gyrus and the inferior sector of the precentral gyrus (five subjects). The overlap tapered both into more anterior prefrontal regions, and posteriorly into the supramarginal gyrus.

Functional Imaging (PET) Studies

Petersen, Raichle, and colleagues have reported several functional imaging (PET) studies in which subjects were required to generate verbs (action words), a paradigm known as "verb generate" (Petersen, Fox, Posner, Mintun, & Raichle, 1988, 1989; Raichle et al., 1994). We have recently conducted a detailed PET study that replicates and extends the findings from the Raichle-Petersen group (Grabowski et al., 1996). We studied 18 normal right-handed volunteers, who underwent both a 3-D MRI study and a PET study, using the same methods described earlier (see first section on Functional Imaging Study). The subjects received injections of $[^{15}O]H_2O$ while performing the verb generate task. In the task, subjects were presented common nouns visually at the rate of one word per 2 sec, and for each, the subjects had to generate (and speak) a verb that went with the noun. Changes in rCBF associated with the task were analyzed using the same methods described earlier (see section on functional imaging; see also H. Damasio et al., 1996).

The strongest areas of rCBF increase associated with the verb-generate task were the left inferior frontal gyrus, left dorsolateral prefrontal cortex, and right cerebellum. These regions correspond closely to those reported in other PET studies using the same paradigm (Petersen et al., 1988, 1989; Raichle et al., 1994; see also Martin et al., 1995). Most importantly, the finding of rCBF increases in the left premotor/prefrontal region during the PET studies is quite congruent with the findings from the lesion studies, which, as reviewed above, have indicated that damage to this region is associated with impaired retrieval of words for actions. Convergent evidence is also available from functional studies using other imaging modalities, such as functional MRI (Hinke et al., 1993).

Conclusions

In sum, we have replicated the initial observation that a deficit in retrieval of words for actions is accompanied by damage in the dorsolateral sector of the left frontal lobe, mostly in the inferior frontal gyrus. Functional imaging studies, both from our laboratory and from others, are consistent with this finding.

CONCLUDING REMARKS

The findings described herein offer support for the idea that, in addition to the separation of the neural systems that support the retrieval of concepts for entities belonging to varied categories, there is a parallel regionalization for the systems that support the retrieval of the word forms corresponding to those entities.

As alluded to earlier, we do not see the neural sites identified in our study as "centers" holding explicit records of word forms, but rather as clusters of neuron ensembles that hold dispositional records for the transient reconstruction of word forms, in appropriate sensory and motor structures, when the processing of the concepts of the corresponding entities activates those ensembles. Some of these neuron ensembles operate as intermediaries between the neural sites that subserve conceptual structure, and the neural sites in which a word form can be transiently reconstructed. We also believe this third-party intermediary role can operate in reverse: to link a word form we may hear or read with the corresponding concept. Elsewhere, we have proposed that these neuron ensembles (convergence–divergence zones) interact dynamically and probabilistically with other regions of cerebral cortex, by means of feedforward and feedback projections (Damasio, 1989a, Damasio & Damasio, 1994). We predict that given similar environmental conditions, normal individuals will develop the type of large-scale system architecture we have uncovered in these studies, although there is ample room for individual variation within each key region. It is important to note, however, that we are not suggesting that there is a rigid brain module out of which one names all persons, animals, and so forth, but rather that, in general, the neural sites identified here will tend to be required when we process words for many though by no means all persons, animals, tools, actions, and so on.

ACKNOWLEDGMENTS

Supported by National Institute of Neurological Disorders and Stroke Program Project Grant NS 19632, and The Mathers Foundation.

REFERENCES

Albert, M. S., Butters, N., & Levin, J. A. (1979). Temporal gradients in the retrograde amnesia of patients with alcoholic Korsakoff's disease. _Archives of Neurology, 36,_ 211–216.

Basso, A., Capitani, E., & Laiacona, M. (1988). Progressive language impairment without dementia: A case with isolated category-specific semantic defect. _Journal of Neurology, Neurosurgery, and Psychiatry, 51,_ 1201–1207.

Baxter, D. M., & Warrington, E. K. (1985). Category-specific phonological dysgraphia. _Neuropsychologia, 23,_ 653–666.

Caramazza, A., & Hillis, A. (1991). Lexical organization of nouns and verbs in the brain. _Nature, 349,_ 788–790.

Carney, A., & Temple, C. M. (1993). Prosopanomia? A possible category-specific anomia for faces. _Cognitive Neuropsychology, 10,_ 185–195.

Cipolotti, L., McNeil, J. E., & Warrington, E. K. (1993. Spared written naming of proper names: A case report. _Memory, 1,_ 289–331.

Cohen, L., Bolgert, F., Timsit, S., & Chermann, J. F. (1994). Anomia for proper names after left thalamic infarct. _Journal of Neurology, Neurosurgery, and Psychiatry, 57,_ 1283–1284.

Damasio, A. R. (1989a). Time-locked multiregional retroactivation: A systems-level proposal for the neural substrates of recall and recognition. _Cognition, 33,_ 25–62.

Damasio, A. R. (1989b). Concepts in the brain. _Mind and Language, 4,_ 24–28.

Damasio, A. R. (1990). Category-related recognition defects as a clue to the neural substrates of knowledge. _Trends in Neurosciences, 13,_ 95–98.

Damasio, A. R. (1992). Aphasia. _New England Journal of Medicine, 326,_ 531–539.

Damasio, A. R., Brandt, J. P., Tranel, D., & Damasio, H. (1991). Name dropping: Retrieval of proper or common nouns depends on different systems in left temporal cortex. _Society for Neuroscience Abstracts, 17,_ 4.

Damasio, A. R., & Damasio, H. (1992). Brain and language. _Scientific American, 267,_ 88–95.

Damasio, A. R., & Damasio, H. (1994). Cortical systems for retrieval of concrete knowledge: The convergence zone framework. In C. Koch (Ed.), _Large-scale neuronal theories of the brain_ (pp. 61–74). Cambridge, MA: MIT Press.

Damasio, A. R., Damasio, H., Tranel, D., & Brandt, J. P. (1990). The neural regionalization of knowledge access: Preliminary evidence. _Quantitative Biology, 55,_ 1039–1047.

Damasio, A. R., Damasio, H., & Van Hoesen, G. W. (1982). Prosopagnosia: Anatomic basis and behavioral mechanisms. _Neurology, 32,_ 331–341.

Damasio, A. R., Grabowski, T. J., Damasio, H., Tranel, D., Frank, R. J., Spradling, J., Ponto, L. L. B., Watkins, G. L., & Hichwa, R. D. (1995). Separate lexical categories are retrieved from separate systems: A PET activation study. _Society for Neuroscience, 21,_ 1498.

Damasio, A. R., & Tranel, D. (1990). Knowing that "Colorado" goes with "Denver" does not imply knowledge that "Denver" *is* in "Colorado." *Behavioural Brain Research, 40,* 193–200.

Damasio, A. R., & Tranel, D. (1993). Nouns and verbs are retrieved with differently distributed neural systems. *Proceedings of the National Academy of Sciences, 90,* 4957–4960.

Damasio, H., & Damasio, A. R. (1989). *Lesion analysis in neuropsychology.* New York: Oxford University Press.

Damasio, H., & Frank, R. (1992). Three-dimensional in vivo mapping of brain lesions in humans. *Archives of Neurology, 49,* 137–143.

Damasio, H., Grabowski, T. J., Tranel, D., Hichwa, R. D., & Damasio, A. R. (1996). A neural basis for lexical retrieval. *Nature, 380,* 499–505.

Daniele, A., Giustolisi, L., Silveri, M. C., Colosimo, C., & Gainotti, G. (1994). Evidence for a possible neuroanatomical basis for lexical processing of nouns and verbs. *Neuropsychologia, 32,* 1325–1341.

Farah, M. J., & Wallace, M. A. (1992). Semantically-bounded anomia: Implications for the neural implementation of naming. *Neuropsychologia, 30,* 609–621.

Fiez, J. A., & Tranel, D. (1994). Retrieval of lexical and conceptual knowledge related to actions and events. *Abstracts of the 66th annual meeting of the Midwestern Psychological Association,* Chicago, Illinois, p. 16.

Fiez, J. A., & Tranel, D. (1997). Standardized stimuli and procedures for investigating the retrieval of lexical and conceptual knowledge for actions. *Memory and Cognition.*

Flude, B. M., Ellis, A. W., & Kay, J. (1989). Face processing and name retrieval in an anomic aphasic: Names are stored separately from semantic information about familiar people. *Brain and Cognition, 11,* 60–72.

Francis, W. M., & Kucera, H. (1982). *Frequency analysis of English usage: Lexicon and grammar.* Boston: Houghton Mifflin.

Franklin, S., Howard, D., & Patterson, K. (1995). Abstract word anomia. *Cognitive Neuropsychology, 12,* 549–566.

Friston, K. J., Frith, C. D., Liddle, P. F., & Frackowiak, R. S. J. (1991). Comparing functional (PET) images: The assessment of significant change. *Journal of Cerebral Blood Flow and Metabolism, 11,* 690–699.

Goodglass, H., & Budin, C. (1988). Category and modality specific dissociations in word comprehension and concurrent phonological dyslexia. *Neuropsychologia, 26,* 67–78.

Goodglass, H., Christiansen, J. A., & Gallagher, R. E. (1994). Syntactic constructions used by agrammatic speakers: Comparison with conduction aphasics and normals. *Neuropsychology, 8,* 598–613.

Goodglass, H., Wingfield, A., Hyde, M. R., & Theurkauf, J. C. (1986). Category specific dissociations in naming and recognition by aphasic patients. *Cortex, 22,* 87–102.

Grabowski, T. J., Damasio, H., Frank, R., Hichwa, R. D., Boles-Ponto, L. L., & Watkins, G. L. (1995). A new technique for PET slice orientation and MRI–PET coregistration. *Human Brain Mapping, 2,* 123–133.

Grabowski, T. J., Frank, R. J., Brown, C. K., Damasio, H., Boles-Ponto, L. L., Watkins, G. L., & Hichwa, R. D. (1996). Reliability of PET activation across statistical methods, subject groups, and sample sizes. *Human Brain Mapping, 4,* 23–46.

Graff-Radford, N. R., Damasio, A. R., Hyman, B. T., Hart, M. N., Tranel, D., Damasio, H., Van Hoesen, G. W., & Rezai, K. (1990). Progressive aphasia in a patient with Pick's disease: A neuropsychological, radiologic, and anatomic study. *Neurology, 40,* 620–626.

Hart, J., Berndt, R. S., & Caramazza, A. (1985). Category-specific naming deficit following cerebral infarction. *Nature, 316,* 439–440.

Hart, J., & Gordon, B. (1992). Neural subsystems for object knowledge. *Nature, 359,* 60–64.

Hillis, A. E., & Caramazza, A. (1991). Category-specific naming and comprehension impairment: A double dissociation. *Brain, 114,* 2081–2094.

Hillis, A. E., & Caramazza, A. (1995). Representations of grammatical categories of words in the brain. *Journal of Cognitive Neuroscience, 7,* 396–407.

Hinke, R. M., Hu, X., Stillman, A. E., Kim, S.-G., Merkle, H., Salmi, R., & Ugurbil, K. (1993). Functional magnetic resonance imaging of Broca's area during internal speech. *NeuroReport, 4,* 675–678.

Hittmair-Delazer, M., Denes, G., Semenza, C., & Mantovani, M. C. (1994). Anomia for people's names. *Neuropsychologia, 32,* 465–476.

Katz, R. B., & Goodglass, H. (1990). Deep dysphasia: Analysis of a rare type of repetition disorder. *Brain and Language, 39,* 153–185.

Levelt, W. J. M. (1989). *Speaking: From intention to articulation.* Cambridge, MA: MIT Press.

Levelt, W. J. M. (1992). Accessing words in speech production: Stages, processes and representations. *Cognition, 42,* 1–22.

Lucchelli, F., & De Renzi, E. (1992). Proper name anomia. *Cortex, 28,* 221–230.

Marin, D. S. M., Saffran, E. M., & Schwartz, M. E. (1976). Dissociations of language in aphasia: Implications for normal function. *Annals of the New York Academy of Science, 280,* 868–889.

Martin, A., Haxby, J. V., Lalonde, F. M., Wiggs, C. L., Ungerleider, L. G. (1995). Discrete cortical regions associated with knowledge of color and knowledge of action. *Science, 270,* 102–105.

McCarthy, R. A., & Warrington, E. K. (1985). Category specificity in an agrammatic patient: The relative impairment of verb retrieval and comprehension. *Neuropsychologia, 23,* 709–727.

McCarthy, R. A., & Warrington, E. K. (1988). Evidence for modality-specific meaning systems in the brain. *Nature, 334,* 428–430.

McKenna, P., & Parry, R. (1994). Category specificity in the naming of natural and man-made objects: Normative data from adults and children. *Neuropsychological Rehabilitation, 4,* 255–281.

McKenna, P., & Warrington, E. K. (1978). Category-specific naming preservation: A single case study. *Journal of Neurology, Neurosurgery, and Psychiatry, 41,* 571–574.

McKenna, P., & Warrington, E. K. (1980). Testing for nominal dysphasia. *Journal of Neurology, Neurosurgery, and Psychiatry, 43,* 781–788.

McNeil, J. E., Cipolotti, L., & Warrington, E. K. (1994). The accessibility of proper names. *Neuropsychologia, 32,* 193–208.

Miceli, G., Silveri, M. C., Nocentini, U., & Caramazza, A. (1988). Patterns of dissociation in comprehension and production of nouns and verbs. *Aphasiology, 2,* 351–358.

Miceli, G., Silveri, M. G., Villa, G. P., & Caramazza, A. (1984). On the basis for the agrammatic's difficulty in producing main verbs. *Cortex, 20,* 207–220.

Miozzo, A., Soardi, S., & Cappa, S. F. (1994). Pure anomia with spared action naming due to a left temporal lesion. *Neuropsychologia, 32,* 1101–1109.

Petersen, S. E., Fox, P. T., Posner, M. I., Mintun, M., & Raichle, M. E. (1988). Positron emission tomographic studies of the cortical anatomy of single-word processing. *Nature, 331,* 585–589.

Petersen, S. E., Fox, P. T., Posner, M. I., Mintun, M., & Raichle, M. E. (1989). Positron emission tomographic studies of the processing of single words. *Journal of Cognitive Neuroscience, 1,* 153–170.

Pietrini, V., Nertempi, P., Vaglia, A., Revello, H. G., Pinna, V., & Ferro-Milona, F. (1988). Recovery from herpes simplex encephalitis: Selective impairment of specific semantic categories with neuroradiological correlation. *Journal of Neurology, Neurosurgery, and Psychiatry, 51,* 1284–1293.

Raichle, M. E., Fiez, J. A., Videen, T. O., MacLeod, A.-M. K., Pardo, J. V., Fox, P. T., & Petersen, S. E. (1994). Practice-related changes in human brain functional anatomy during nonmotor learning. *Cerebral Cortex, 4,* 8–26.

Riddoch, M. J., & Humphreys, G. W. (1987). Visual object processing in optic aphasia: A case of semantic access agnosia. *Cognitive Neuropsychology, 4,* 131–185.

Roelofs, A. (1993). Testing a non-decompositional theory of lemma retrieval in speaking: Retrieval of verbs. *Cognition, 47,* 59–87.

Rosch, E., Mervis, C. B., Gray, W. D., Johnson, D. M., & Boyes-Braem, P. (1976). Basic objects in natural categories. *Cognitive Psychology, 8,* 382–439.

Sacchett, C., & Humphreys, G. W. (1992). Calling a squirrel a squirrel but a canoe a wigwam: A category-specific deficit for artefactual objects and body parts. *Cognitive Neuropsychology, 9,* 73–86.

Sartori, G., & Job, R. (1988). The oyster with four legs: A neuropsychological study on the interaction of visual and semantic information. *Cognitive Neuropsychology, 5,* 105–132.

Semenza, C., & Sgaramella, T. (1993). Proper names production: A clinical case study of the effect of phonemic cueing. *Memory, 1,* 265–280.

Semenza, C., & Zettin, M. (1988). Generating proper names: A case of selective inability. *Cognitive Neuropsychology, 5,* 711–721.

Semenza, C., & Zettin, M. (1989). Evidence from aphasia for the role of proper names as pure referring expressions. *Nature, 342,* 678–679.

Shallice, T., & Kartsounis, L. D. (1993). Selective impairment of retrieving people's names: A category specific disorder? *Cortex, 29,* 281–291.

Silveri, M. C., & Gainotti, G. (1988). Interaction between vision and language in category-specific semantic impairment. *Cognitive Neuropsychology, 5,* 677–709.

Small, S. L., Hart, J., Nguyen, T., & Gordon, B. (1995). Distributed representations of semantic knowledge in the brain. *Brain, 118,* 441–453.

Snodgrass, J. G., & Vanderwart, M. (1980). A standardized set of 260 pictures: Norms for name agreement, image agreement, familiarity, and visual complexity. *Journal of Experimental Psychology: Human Learning and Memory, 6,* 174–215.

Talairach, J., & Szikla, G. (1967). *Atlas d'anatomie sterotaxique du telencephale.* Paris: Masson et Cie.

Tranel, D. (1991). Dissociated verbal and nonverbal retrieval and learning following left anterior temporal damage. *Brain and Cognition, 15,* 187–200.

Tranel, D., Damasio, H., & Damasio, A. R. (1995). Double dissociation between overt and covert face recognition. *Journal of Cognitive Neuroscience, 7,* 425–432.

Tranel, D., Damasio, H., Damasio, A. R., & Brandt, J. P. (1995). Separate concepts are retrieved from separate neural systems: Neuroanatomical and neuropsychological double dissociations. *Society for Neuroscience, 21,* 1497.

Warrington, E. K. (1975). The selective impairment of semantic memory. *Quarterly Journal of Experimental Psychology, 27,* 635–657.

Warrington, E. K. (1981). Concrete word dyslexia. *British Journal of Psychology, 72,* 175–196.

Warrington, E. K., & McCarthy, R. A. (1983). Category specific access dysphasia. *Brain, 106,* 859–878.

Warrington, E. K., & McCarthy, R. A. (1987). Categories of knowledge: Further fractionations and an attempted integration. *Brain, 110,* 1273–1296.

Warrington, E. K., & Shallice, T. (1984). Category specific semantic impairments. *Brain, 107,* 829–854.

Woods, R. P., Mazziotta, J. C., & Cherry, S. R. (1993). MRI–PET registration with automated algorithm. *Journal of Computer Assisted Tomography, 17*, 536–546.

Zingeser, L. B., & Berndt, R. S. (1988). Grammatical class and context effects in a case of pure anomia: Implications for models of language production. *Cognitive Neuropsychology, 5*, 473–516.

Zingeser, L. B., & Berndt, R. S. (1990). Retrieval of noun s and verbs in agrammatism and anomia. *Brain and Language, 39*, 14–32.

Part III

Dissociations and Other Naming Phenomena

The internal structure of the naming process has been illuminated by recent studies of the fractionation of object naming capacities along the lines of input and output channels, on one hand, and along the lines of the concepts to be named, on the other. Freund's (1889) description of 'optic aphasia' first called attention to the need to provide for a gateway between relatively peripheral, prelinguistic input or output systems and a more central system of knowledge representation and name access. In the first of the two chapters in this section, Ria De Bleser presents a synthesis of single-channel deficits of naming that may affect particular input or output pathways.

Section II includes Tranel, Damasio, and Damasio's treatment of the anatomical basis for a number of category-specific naming dissociations, including those affecting the retrieval of proper nouns. In Chapter 5 of Part III, Carlo Semenza elaborates on the cognitive implications of aphasia for proper names, on the basis of in-depth study of his own clinical cases, and their relation to other cases in the literature.

Chapter 4

Modality-Specific Lexical Dissociations

R. De Bleser

INTRODUCTION

The successful naming of visually presented stimuli such as objects, pictures, or written words depends on intact linguistic as well as visual-perceptual processes. These may be selectively affected after brain damage. Following a focal lesion in the left perisylvian area, there is generally an impairment of language (aphasia) without concomitant perceptual disorders, whereas occipital lesions may lead to impairments of visual recognition (agnosia) without associated verbal or other nonverbal disorders. Selective aphasias and agnosias have also been observed in cases of slowly progressive degenerative disease ("probable Alzheimer's disease"), where the progression may generalize to other cognitive functions as late as 11 years postonset (Becker, Huff, Nebes, Holland, & Boller, 1988; De Renzi, 1986). Such selective patterns of impaired language versus spared visual recognition and vice versa are taken as strong evidence for the modular independence of these systems. Accordingly, language as well as perception are highly specialized and dissociable modules in the human cognitive system (Marr, 1982; Marshall, 1989).

However, there is still controversy about the internal architecture of the linguistic and perceptual processing systems and their interaction in naming, and this is the topic of the present chapter. With respect to naming a visual word in reading, the current majority view is that there are multiple routes from recognition to name output (e.g., Shallice, 1988). One route, which is called "direct," involves attaching a spoken label immediately to the graphemically recognized word. Another, indirect route requires mediation via the semantic system. With respect to object naming, a direct route from object recognition to name output has also been proposed by some authors (e.g., Shuren, Geldmacher, & Heilman, 1993), but most

ANOMIA: Neuroanatomical and Cognitive Correlates

93

authors hold the view that the articulation of a response to an object is always mediated by semantics or the "cognitive system" (Seymour, 1979). The nature of the semantic system itself is also controversial (Job & Sartori, 1988), in particular whether it is unitary (e.g., Rapp, Hillis, & Caramazza, 1993) or whether it consists of multiple sensory-based knowledge systems (tactile, visual, gustatory, etc.) to be distinguished from a suprasensory verbal semantic system (e.g., Shallice, 1993).

The contemporary models of visual perception, language, and their interface are partly derived from classical aphasiology, but they differ in important ways. In the next section, we will clarify the inheritance from 19th-century aphasiologists.

FROM MODALITY-SPECIFIC TO SUPRAMODAL

In the classical era of German aphasiology at the end of last century, the first models of language and linguistic disorders were concerned with the processing of single words. The word was considered to be a linguistic sign arbitrarily relating a sound structure with a meaning representation, which Saussure (1916/1959) would later call *"signifigant"* (signifier) and *"signifié"* (signified). Following C. Wernicke (1874, 1886, 1906), some aphasiological authors (e.g., Kleist, 1914, 1934) fractionated the sound structure of words into two distinct modality-specific representational stores (see Figure 1), **a/A** for the auditory representation of words (cf. the phonological

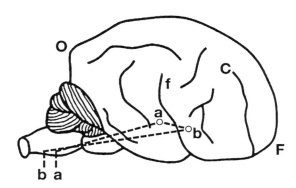

Figure 1 Classical dual lexicon model after Wernicke (1874). b (M in other authors) in Broca's area is a store of the phonological output representations of words for speech production; a (A in other authors) in Wernicke's area stores the phonological input representations of words for speech perception.

input lexicon in contemporary models) and **b/M,** for the articulatory word stores (cf. the phonological output lexicon). Each primary phonological component could be selectively impaired resulting in a particular cardinal symptom, disturbed comprehension for a disorder in **A,** disturbed production for a disorder in **M.** Furthermore, as a consequence of this selective impairment, associated performance deficits were predicted in different tasks such as repetition, reading aloud, writing to dictation, etc. The set of symptoms associated with a disorder in **A** was defined as the syndrome of Wernicke's aphasia, with an impairment in **M** as Broca's aphasia. With a lesion of both, there was global aphasia. Furthermore, each component had its corresponding brain hardware, the superior temporal lobe for **A,** the third frontal convolution for **M.**

An alternative classical theory did not fractionate the sound structure of language into two equivalent representational systems (A and M). Authors in this tradition assumed a central phonological system ("inner speech") which was auditorily based (e.g., Freud, 1891/1935; Kussmaul, 1877). Some of these authors still granted representational functions (motor memories) to the center **M,** but access to it depended on an intact center **A** in the temporal lobe; for others, **M** was merely an articulatory coordination organ.

There was also considerable debate in classical aphasiology concerning theoretical models of object naming. In particular, there was a dispute between authors adopting a single suprasensory semantic system and those who favored multiple sensory-specific systems (optic, tactile, gustatory, etc. semantic memories). The first model of object naming in classical German aphasiology (Kussmaul, 1877) proposed a single concept center for an object. Because in this model there was no direct link between a concept and the output systems, naming had to be mediated by the auditory word form. Comprehension disorders should thus invariably accompany an impairment of naming. Charcot (1883a,b) introduced two important changes to Kussmaul's model. First, he argued from the dissociable memories of his famous patient H that there are sensory-specific memory centers apart from the central conceptual system. Second, he proposed direct links between concepts and each of the motor output systems, for speech as well as for writing, so that there could be pure modality-specific oral or written naming disorders without impaired comprehension.

Wernicke (1886) essentially adopted Charcot's proposal of multiple memory systems, but he rejected a separate suprasensory concept center. In his view, there were only multiple sensory memory images localized in the proximity of the sensory reception areas involved in their development. The "object-center" (**B** for "Begriffe" or Concepts) was regarded as but an artificial abbreviation for the total association of these memory images (see Figure 2). These images were connected to the auditory as well as the motor word representations.

Because the primary sensations arose at different loci in the cerebral cortex, Wernicke assumed that the sensory memories of objects based on them were also spread out over the entire hemispheres. Object naming and voluntary speech in general required that the associated bilaterally represented sensory object memories (the object concept **B**) could access the motor phonological output lexicon **M** in the left hemisphere. Meaningful comprehension and repetition of words was based on a link between the left temporal auditory phonological input lexicon **A** and the bilateral object concepts in **B**. In general, because there was no unitary meaning center, there was no provision for a cortical syndrome of anomic aphasia, a focal supramodal naming disorder, and only diffuse brain damage would give rise to a functional lesion of the postulated intersensory association network.

Wernicke's multiple semantics model of naming led to the first observations of unimodal aphasias. These are disorders of naming without demon-

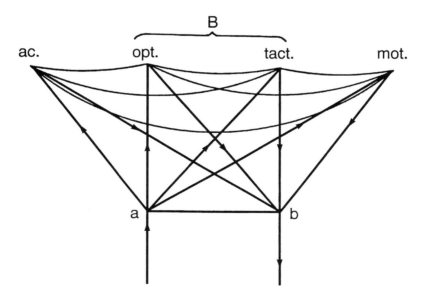

Figure 2   Model with multiple semantic systems (acoustic, optic, tactile, motor) in Wernicke (1886). a = A in other authors, b = M.

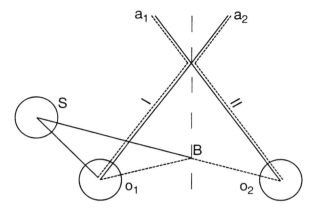

Figure 3 Model of optic aphasia. a_1, a_2; retinae; o_1, o_2: visual cortex. S, speech area; B, splenium. (After Freund, 1889.)

strable recognition impairment restricted to object presentations in one sensory modality, for example, with visual but not with tactile or verbal presentation. Freund (1888, 1889), working in Wernicke's clinic, reported his first observations of optic aphasia and descriptions of tactile aphasia were soon to follow (Redlich, 1894, in Wolff, 1904). Also following in Wernicke's tradition, Lissauer (1890) described a patient with a visual modality-specific recognition problem for an object that also resulted in a unimodal object naming disorder. He called the impairment an associative visual agnosia, which he distinguished from apperceptive agnosia, a visual-recognition disorder already affecting earlier perceptual stages.

Freund interpreted cases of unimodal aphasia as a disorder of association between a single sensory memory system (e.g., the visual store) and the speech system (see Figure 3). Different types of optic aphasia were predicted depending on the particular type of lesion or disconnection involved. Optic aphasia could be the result of a combined intrahemispheric disconnection between an undamaged O_1 and S in the left hemisphere together with an interhemispheric disconnection between an undamaged right hemispheric O_2 and S. In this case, the language system as well as the visual object memories would be fully preserved. Damage to the left occipital region (O_1) together with a lesion of the splenium (B) would give rise to partial object agnosia, namely, for those objects that had stored representations in the left hemisphere only, whereas there would be optic aphasia for those objects that were represented in the right hemisphere. The patient recognizes these objects with O_2; he can demonstrate their use, but he cannot name them since the pathway linking O_2 and the left hemisphere speech area (S) is interrupted in B. Similarly, there could be optic aphasia for objects stored in the left hemisphere as a result of an intra-hemispheric disconnection between an undamaged O_1 and S, and the lesion of O_2 in the right hemisphere would give rise to partial visual agnosia (i.e., a disorder of recognition), namely, for those objects that were represented exclusively in O_2. Complete visual agnosia would result from a lesion of both O_1 and O_2 (cortical agnosia) or if the optic memory centers were disconnected from the other (tactile, gustatory, etc.) sensory memory centers (transcortical agnosia).

Freund did not specify whether the visuo-verbal disconnection giving rise to optic aphasia was to the motor or auditory word form of the speech area S. However, other authors argued for one or the other. For the authors strictly following Wernicke, unimodal aphasias were transcortical motor disorders disconnecting a sensory memory system from the motor representation of the word. In the alternative view, they were transcortical sensory disorders that reflected disconnections between sensory object representations and the auditory word image.

Whereas object agnosia was quite generally accepted, optic aphasia was dismissed as a form of visual agnosia (Freud, 1891/1935). Wolff (1904) rejected the published cases of optic aphasia and tactile aphasia as well as the hypothetical acoustic, gustatory, and olfactory aphasias. His criticism rightly addressed the poor quality of the evidence, but more importantly, Wolff and other critics to follow (e.g., Lange, 1936) disagreed with the assumption of independent sensory-based object memories to which the observations of unimodal aphasias referred. To them, optic and tactile aphasias were phenomena to be subsumed within a more general naming disorder or, at best, to be considered as mild forms of recognition impairments (i.e., object agnosia).

At the end of the classical era of German aphasiology, there was almost general consensus that object naming disorders in aphasia were always supra- or multimodal, and that word-naming impairments in reading were invariably caused by aphasia. Only visual object-recognition disorders or letter-recognition disorders could lead to unimodal impairments in visual agnosia or literal alexia. Some authors (e.g., Freund, 1889) argued that these would always co-occur, because letters did not have a special optic memory status in the theory of the Wernicke school.

A RETURN TO MODALITY SPECIFICITY

Careful analyses of the reading and writing of individual brain-damaged patients showed a variety of modality-specific impairments that could not be reconciled with the classical and neo-classical views. In particular, different forms of subtotal impairments of reading (dyslexia) and writing (dysgraphia) were described without a comparable disorder of spoken language. For example, some patients made semantic errors in reading but not in speech (e.g., Coltheart, Patterson, & Marshall, 1980); others could read orthographically regular but not irregular words (Patterson, Marshall, & Coltheart, 1985); some patients had better preserved written naming than oral naming of pictures (Bub & Kertesz, 1982; Hier & Mohr, 1977). The wealth of cases with such modality-specific effects seemed to provide sufficient evidence that the language processor must have specialized components and routines for graphemic and phonological word forms, as suggested in contemporary models of graphemic processing.

Also with respect to object naming, new cases of optic aphasia (Beauvois, 1982; Lhermitte & Beauvois, 1973), tactile aphasia (Beauvois, Saillant, Meininger, & Lhermitte, 1978), and acoustic aphasia (Denes & Semenza, 1975) were reported that deviated from the generally assumed pattern that in the absence of recognition problems, naming impairments are multimodal, independent of the sensory input modality (Goodglass, Barton, & Kaplan, 1968). As in the classical era, these cases were taken to support claims that there are separate knowledge systems concerned with visual, tactile, and auditory representations.

Such observations on modality-specific impairments of word and object processing fostered a great increase in the use of in-depth single case studies. These became a topic of major theoretical interest again in contemporary cognitive neuropsychology, because they could provide evidence for classical double dissociations between tasks and could thus be used to test the architecture of information-processing models (Shallice, 1988) in a way that was not possible in group studies. If patient X has preserved performance on task A addressing a functional component M in the model but is severely impaired on task B, which is designed to test component N, and if patient Y shows the reverse pattern, with normal performance on task B and a severely impaired performance on task A, then the two tasks are doubly dissociated, and there is evidence for the functional independence of the two components as was assumed in the system.

One of the leading models for normal word processing, the logogen model (see Figure 4), was extensively used as a reference frame in cognitive neuropsychology (e.g., Ellis, 1984; Morton, 1979, 1980a,b; Newcombe & Marshall, 1980; Patterson, 1988). The logogen model and similar dual-route models make a distinction between receptive and productive word stores (but see, for example, Allport, 1984, for an alternative view), and it assumes that written language does not depend on spoken language but

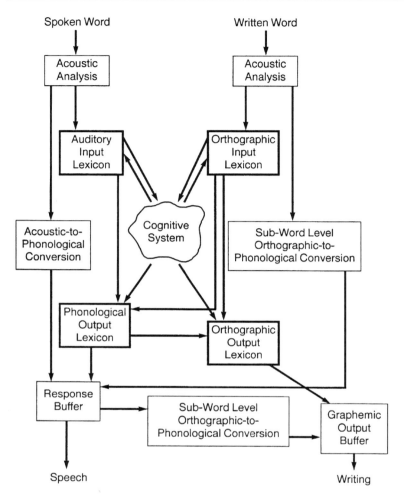

Figure 4 Logogen model of word processing. (After Morton & Patterson, 1980, for explanation, see text.)

that it is processed in a functionally independent system. Four lexicons are thus distinguished which contain only word forms that can be activated without their corresponding meaning, as in word reading without comprehension. Each lexicon can be specifically impaired while word activation in the undisturbed lexicons remains possible. The semantic system is taken to be unitary, nonmodality-specific, and it contains meaning structures without their corresponding word forms. The connections between the lexicons and the semantic system allow word forms to be mapped onto their meaning as is required, for example, in auditory or visual word-to-picture matching. In addition to the lexical word-processing systems, nonlexical systems are adopted for the segmental processing of auditory and graphemic stimuli. As nonwords do not have a lexical entry, they can only be processed in this way. Lexical processing of words as well as nonlexical processing of nonwords or words is based on prelexical pattern recognition for the perception, identification, and categorization of linguistic elements. The modality-specific analysis systems allow, for example, auditory or visual discrimination of stimuli.

Unimodal impairments of picture naming have also been interpreted within the standard logogen model and provided evidence for modality-specific word-form output lexicons. For example, Bub and Kertesz (1982) described a patient who was much more impaired in oral naming than in written naming. The patient thus had impaired access from semantics to the phonological but not to the graphemic output lexicon. Further cases supporting a modality-specific lexical distinction showed a frequency effect in either oral or written picture naming, indicating an impairment to one of the output lexicons rather than to the semantic system. Kay and Ellis (1987), for example, reported a patient who showed such a frequency effect in oral but not in written naming, so that the deficit had to be localized at the level of the phonological output lexicon. Similarly, Goodman and Caramazza (1986) reported a patient who made errors on writing allographs (e.g., tow, toe) dictated together with a disambiguating definition. Because the semantically inappropriate allographic word was produced predominantly if this was of higher frequency than the target, there seemed to be evidence for a selective impairment of the graphemic output lexicon.

The frequently observed compatibility of dysphasic, dyslexic, and dysgraphic error patterns with models of normal processing led to the success of modern cognitive neuropsychology. On the one hand, impairments of word processing could obviously be interpreted as specific lesions to one or more functional components of models of normal processing; on the other hand, the detailed assessment of word-processing disorders in single cases allowed further extensions and modifications of these models. Especially with respect to the semantic system, the discussion has not been settled whether a further fractionation into subsystems for different sensory modalities is theoretically desirable and necessary for the explanation of modality-specific anomias. The controversy is reflected in the titles of recent articles, such as "The multiple semantics hypothesis: multiple confusions?" (Caramazza, Hillis, Rapp, & Romani, 1990) versus "Multiple semantics: Whose confusions?" (Shallice, 1993). Furthermore, it is not clear whether object naming—in contrast to word naming—is always mediated by semantics (Seymour, 1979) or whether a direct route should also be postulated from object recognition to the phonological or graphemic output lexicon while bypassing semantics (Kremin, 1986; Shuren et al., 1993). As in the classical days, the discussion is centered around the existence of modality-specific object anomias and their difference from object agnosias.

What needs to be accounted for is that in unimodal anomia such as optic aphasia, there is a naming impairment specific to visually presented objects. Stimuli presented in the tactile or auditory modality can be named very well, and the patient is able to provide the name corresponding to a verbal description. Thus, the linguistic system seems to be preserved and sufficient semantics can apparently be activated to retrieve a phonological address from nonvisual input. Furthermore, the patients do not show any obvious recognition problems either, in contrast to object agnosic patients. They can draw from memory objects that they have just seen but could not name, and they can also do object decision tasks when the false picture is constructed by replacing parts of an object by parts of another object. Finally, optic aphasic patients even seem to have access to some semantic knowledge from visually presented objects, given that they can indicate by means of gestures the use of objects they cannot name, categorize objects into the appropriate class, or identify the object in word-to-picture matching tasks.

The interpretation of such cases of optic (or tactile or auditory) aphasia seemed to be impossible within a unitary semantics, single indirect route framework such as the pictogen model (see Figure 5). The standard logogen–pictogen system has a three-stage model of naming visually presented objects and does not provide for distinctions according to sensory modality. The first stage in the model concerns the recovery of distal information in the formation of a temporary representation with a viewpoint-dependent object description. This stage is also called visual analysis. In the second stage, corresponding representations within the "cognitive system" are activated. This may be divided into representations that concern object recognition (so-called stored structural descriptions) and those concerning semantic knowledge. Stored object-centered structural descriptions are also called pictogens analogous to the logogens for word recognition. A pictogen is activated by all views of the same object and can fire to visually very dissimilar versions of the same object type (Warren & Morton, 1982). In a third stage, a phonological or graphemic entry appropriate to the semantic representation is activated in one of the output lexicons.

Within this framework, object agnosia could easily be interpreted as a problem at the level of the pictogens or access to them. This would explain

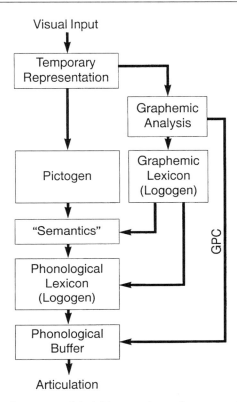

Figure 5 Pictogen model of object naming: unitary semantics and a single indirect route to name output. GPC, Grapheme–phoneme correspondence. (After Morton, 1985.)

the modality specificity of the naming impairment as a result of a visual-recognition problem. However, the model seemed to be inadequate to account for optic aphasia. The impairment cannot be localized at the pictogen, because there is intact recognition, nor can it be localized at semantics, given that naming is possible from other sensory modalities. A disconnection of semantics from the pictogens cannot explain that patients have semantic knowledge of the objects they cannot name. It was therefore suggested that providing the appropriate gesture to an unnamed object is not indicative of semantic access, and that so-called optic aphasics would have a mild though often unassessed recognition impairment. As in the classical days, then, the existence of modality-specific aphasias was denied and they were seen as milder forms of object agnosia (e.g., Bauer & Rubens, 1985).

The ongoing discussion on optic aphasia, its distinction from object agnosia, and the relevance it may have for competing models of semantics has to do among others with difficulties in assessing the intact status of recognition. In a recent case study, Davidoff and De Bleser (1994) showed that recognition ability can be easily overestimated. The patient, HG, was originally thought to have optic aphasia because her picture naming was poor, although tactile and verbal description naming was normal. She seemed to have good visual recognition, because she did not show any difficulties in matching atypical to typical object views, and she could sort line drawings into real versus unreal objects. She could also successfully categorize line drawings into fine-grained classes such as domestic versus foreign animals. However, HG lacked an object superiority effect on a tachistoscopically presented test for matching a chair or a "scrambled chair" to one of two minimally different versions (Davidoff & Donnelly, 1990). Thus, HG's recognition was experimentally shown to be subnormal, and this was also indicated by the following observations. The patient's naming greatly improved if additional perceptual information was provided, so that she named visually presented real objects significantly better than photographs or line drawings of the same objects. Her naming of photographs of objects taken from a typical view was also much better than from atypical views. Furthermore, errors in naming were not semantic but they showed a visual similarity to the target. Thus, there was a recognition impairment underlying HG's apparent optic aphasia, in other words, she had a form of object agnosia, and the authors argued that HG's good performance on a variety of "recognition" tests from a damaged recognition system was obviously due to the fact that is possible to activate some object knowledge without complete recognition by an analysis of the object parts (Hillis, Rapp, Romani, & Caramazza, 1990; Warrington & James, 1988).

Davidoff and De Bleser (1993) then surveyed the literature on patients classified as having either object agnosia or optic aphasia using two features of HG's performance, namely, the influence of stimulus characteristics on naming and the near absence of semantic errors. The published cases who were variously called visual agnosic or optic aphasic could be divided into two groups. There were 15 patients described in sufficient detail who, like HG, were significantly better in naming real objects than pictures of objects and who, also like HG, did not produce a substantial proportion of semantic errors (see Table 1). The influence of stimulus characteristics on naming in group 1 patients and the absence of semantic errors suggested that the source of the modality-specific anomia in these patients was in recognition. Another group of 9 patients did not show an influence of stimulus characteristics on naming and they produced predominantly semantic errors or circumlocutions (see Table 2). Group 2 patients, then, showed a pattern one would expect in optic aphasia that could

Table 1 Modality-Specific Anomia with Effect of Stimulus Quality

	Real objects (%)	Picture/object difference	Error types
Larrabee, Levin, Huff, Kay, & Guinto (1985) CE	100	Yes	Not semantic
Damasio, McKee, & Damasio (1979) Case 2	100	Yes	Not semantic
Damasio et al. (1979) Case 1	92	Yes	Not semantic
Oxbury, Oxbury, & Humphrey (1969) Case 1	90	Yes	Many types
Iorio, Falanga, Fragassi, & Grossi (1992)	80	Yes	Visual/semantic
Beauvois & Saillant (1985) RV	Good	Yes	Visual
Riddoch & Humphreys (1987b)	65	Yes	Visual
Mack & Boller (1977)	65	Yes	Visual
Newcombe & Ratcliff (1974)	40	Yes	Visual
Levine (1978)	Variable	Yes	Visual
Kertesz (1979)	37	Yes	Visual/perseveration
Kawahata & Nagata (1989)	36	yes	Visual
Milner et al. (1991)	35	Yes	Visual
Grailet, Seron, Bruyer, Cayette, & Frederix (1990)	27	Yes	Visual
Lhermitte, Chedru, & Chain (1973)	12	Yes	Visual

Table 2 Modality-Specific Anomia without Effect of Stimulus Quality

	Real objects (%)	Picture/ object difference	Error types
Pena-Casanova, Roig-Rovira, Bermudez, & Tolosa-Sarro (1985)	72	No	Paraphrase
Lhermitte & Beauvois (1973)	70	No	Semantic
Oxbury et al. (1969) Case 2	65	No	Paraphrase
Gil et al. (1985)	64	No	Semantic
Hécaen, Goldblum, Masure, & Ramier (1974)	55	No	Many types
Caplan & Hedley-White (1974)	50	No	Semantic
Riddoch & Humphreys (1987a)	46	No	Visual + semantic
Lindeboom & Savinkels (1986)	44	No	Semantic
Larrabee et al. (1985) WM	30	No	Semantic

not be accounted for by models of unitary semantics, which provide a single indirect route to name output, like the pictogen model in Figure 5.

A variant of this approach also assumes a unitary semantic system but one that is represented in duplicate in the left and right hemisphere (Coslett & Saffran, 1989). Optic aphasia would then arise if the left hemisphere can no longer process visual stimuli due to a left occipital lobe infarct, and the preserved visual and semantic processing of the right hemisphere cannot access left-hemisphere phonology due to a callosal lesion. This explains that optic aphasia may be restricted to an optic *anomia*, and that semantic comprehension, assumed to be bilaterally represented, may be relatively preserved. The authors explicitly refer to Freund (1889) as their historical precedent. The approach differs from Freund, however, in the assumption of a verbal or suprasensory semantic system. Freund and more generally the Wernicke school to which he belonged did not assume such a suprasensory system. Furthermore, Freund makes a provision for representationally dissimilar left- and right-hemisphere visual semantics.

Although the duplicate unitary semantics model may explain the pattern of some patients with optic aphasia, it cannot account for patients who are able to process visual stimuli in the left hemisphere, yet are unable to name them.

An alternative account within a unitary semantics framework assumes a dual-route framework for objective naming in analogy to the dual-route model of word naming (see Figure 6). It is proposed that in optic aphasia the direct route from visual recognition to phonology is impaired, so that patients have to rely exclusively on the indirect route via semantics. This by itself is insufficient and too imprecise to produce the correct name output (Ratcliff & Newcombe, 1982). Within this framework, optic aphasia has recently been contrasted to its mirror image, so-called "nonoptic" aphasia. In "nonoptic aphasics," visual-object naming is remarkably better than the performance on semantic tasks would lead one to expect, so that naming seems to proceed via a direct route from visual recognition to name output while bypassing semantics (Shuren et al., 1993). However, it has been objected (Shallice, 1993) that a mere impairment of a direct route either for word or object naming cannot explain the occurrence of semantic errors, but that an additional semantic deficit must be postulated, which again raises the problem of how the semantic system should be modeled.

The most influential and straightforward account of optic aphasia in modern neuropsychology offers an explanation within a multiple semantics–single route model (Beauvois, 1982). In the so-called representational account of multiple semantics (Riddoch, Humphreys, Coltheart, & Funnell, 1988), modality-specific semantic systems are postulated which contain information specific to a particular sensory modality, so-called senso-

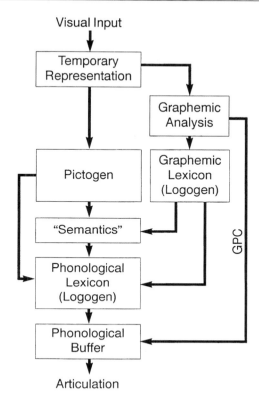

Figure 6 Dual route model of object naming: unitary semantics and a direct as well as an indirect route to name output. GPC, Grapheme–phoneme correspondence. (After Newcombe & Radcliff, 1974.)

ry knowledge. This feeds into a suprasensory verbal semantic system containing functional and associative knowledge about the object (note that this assumption also deviates from Freund's original view). In contrast to the direct route model, access to phonology is assumed to be possible only from the verbal system, not directly from the modality-specific semantic systems (see Figure 7).

The symptomatology of optic aphasia is then interpreted as a visuo-verbal disconnection. The verbal semantic system can be accessed from non-visual modality-specific semantic systems and can retrieve the appropriate name in output phonology for tactile, auditory, and so forth, naming. The visual semantic system by itself functions sufficiently to allow the production of adequate gestures to visual objects or access sensory semantic

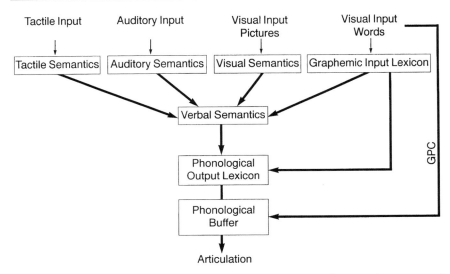

Figure 7   Multiple semantics model of object naming: sensory-specific semantic systems and a single indirect route to name output. GPC, Grapheme–phoneme correspondence. (After Shallice, 1988.)

knowledge in general. However, a disconnection between the preserved visual and verbal semantic systems results in optic anomia, because names can only be retrieved from the verbal system, and this has become inaccessible from vision. The error types will of necessity be semantic, even if the disconnection is complete, since they result from the visual semantic system.

The unitary semantics framework can explain the presence of semantic errors by assuming only partial disconnection from recognition in the pictogen to the semantic system. Because there is no semantic knowledge in the pictogen, semantic errors cannot arise at this stage, and they must reflect partial access to semantics. This raises the difficulty of distinguishing access from storage problems.

FROM SEQUENTIAL STAGE MODELS TO CONNECTIONIST MODELS OF NAMING

The problem of distinguishing access from storage problems becomes neutralized in connectionist models with unitary semantics. In contrast to sequential stage models, which assume that processing at one stage is complete before processing starts at the next level, connectionist models allow information to "cascade" continuously throughout the network.

For example, in the model proposed by Humphreys, Riddoch, and Quinlan (1988), a picture temporarily activates all similarly looking items at the level of structural descriptions (or pictogens). Each of these structural descriptions activates a semantic representation. Thus, each node in the system activates the corresponding node at the next processing level and inhibits the other nodes at its processing stage. The correct structural description will inhibit all the others, and the correct semantic representation will inhibit all the others. At some point in the process, the semantic representations of objects that are also structurally similar will be particularly highly activated. If optic aphasics have problems selecting the correct node among highly activated nodes, they will produce semantic errors especially to items in semantic categories that also have structural similarity, as was the case for JB reported in Riddoch and Humphreys (1987b) and Humphreys et al. (1988).

Like other optic aphasics, the patient JB was impaired in naming visually presented objects but could show how they were used by means of gestures and perform difficult object decision tasks in which the incorrect items were constructed by within-category replacements, thus demonstrating correct access to structural knowledge from vision. The patient did not seem to have an impairment at the level of the pictogen because he could perform word-to-picture matching tasks if the distractor had a visual similarity to the target. His performance was not affected either by semantic similarity between distractor and target, so that the semantic representations seemed to be spared. However, JB's performance on word–picture matching was significantly worse when there was both visual and semantic similarity between the target and the distractor. This was in line with his naming ability, which was also affected by structural similarity (i.e., when members of a particular semantic category, for example biological items, are also visually related). A similar patient (KR) with a unimodal category-specific impairment for biological items was reported by Hart and Gordon (1992). The cascade framework seemed to provide a relatively straightforward explanation of such patients by suggesting that the pathway between structural information and the unitary system of semantic knowledge is affected.

However, the controversy on unitary versus multiple semantics also entered connectionist approaches to object knowledge and modality-specific (and category-specific) anomia. Farah (1994), for example, has argued that, following Warrington and colleagues, the double dissociation between knowledge of living and nonliving things reported by Warrington and McCarthy (1983, 1987) and Warrington and Shallice (1984) should be interpreted as a modality-specific difference between preserved versus impaired visual-sensory semantics and verbal-functional semantics, respectively, rather than as a difference within a unitary semantic system between the categories living and nonliving. The argument is that our distinctive knowledge of living things such as animals is primarily visual, whereas that of nonliving things is primarily functional. A disorder of structural descriptions in some types of visual object agnosia will thus primarily affect biological items while preserving human-made objects.

Farah and McClelland (1991) built a connectionist network implementing these hypotheses. The authors found that damage to the visual knowledge system indeed mainly affected the recognition of biological items, whereas damage to the semantic system in the network caused a selective deficit for human-made objects. Farah (1994) furthermore argued that a connectionist model can better explain the patient data than a sequential stage model, in that damage to the connectionist system is nonlocal. If visual semantics is impaired, the functional semantics in the undamaged component will also function abnormally and the reverse because of a loss of so-called collateral support between the subsystems.

Given the interactive nature of such spreading activation models and the nonlocal effects of lesions to the model, the existence of pure modality-specific deficits is being questioned once more. As these models do not provide functionally independent components of a system, modality-specific double dissociations can no longer serve as a test of the modular architecture either of language or vision. However, judging from some of the peer commentaries to Farah (1994), modular models and spreading activation connectionist models will continue to exist side-by-side in neuropsychology, and modality-specific lexical impairments will remain of theoretical interest at least until connectionism has met the challenge that "the parameters of these models can be set in such a way that the present set of data as well as other evidence . . . receive a satisfactory answer" (Schriefers, Meyer, & Levelt, 1990, p. 100).

ACKNOWLEDGMENTS

With kind acknowledgments to J. Davidoff, with whom much of the work reported here was either conducted or discussed or both.

REFERENCES

Allport, D. A. (1984). Speech production and comprehension: One lexicon or two? In W. Prinz & A. F. Sanders (Eds.), *Cognition and motor processes* (pp. 209–228). Berlin: Springer.
Bauer, R. M., & Rubens, A. B. (1985). Agnosia. In K. M. Heilman & E. Valenstein (Eds.), *Clinical neuropsychology* (2nd ed.) (pp. 215–278). New York: Oxford University Press.
Beauvois, M. F. (1982). Optic aphasia: A process of interaction between vision and language. *Philosophical Transactions of the Royal Society of London B, 298*, 35–47.

Beauvois, M. F., & Saillant, B. (1985). Optic aphasia for colours and colour agnosia: a distinction between visual and visuo-verbal impairments in the processing of colours. *Cognitive Neuropsychology, 2,* 1–48.

Beauvois, M. F., Saillant, B., Meininger, V., & Lhermitte, F. (1978). Bilateral tactile aphasia: A tacto-verbal dysfunction. *Brain, 101,* 381–401.

Becker, J. T., Huff, F. J., Nebes, R. D., Holland, A., & Boller, F. (1988). Neuropsychological function in Alzheimer's disease: Pattern of impairment and rates of progression. *Archives of Neurology, 45,* 263–268.

Bub, D. N., & Kertesz, A. (1982). Evidence for lexicographic processing in a patient with preserved written over oral single word naming. *Brain, 105,* 697–717.

Caplan, L., & Hedley-White, T. (1974). Cuing and memory dysfunction in alexia without agraphia: Case report. *Brain, 97,* 251–262.

Caramazza, A., Hillis, A. E., Rapp, B. C., & Romani, C. (1990). The multiple semantics hypothesis: Multiple confusions? *Cognitive Neuropsychology, 7,* 161–189.

Charcot, J. M. (1883a). Des différentes formes de l'aphasie—De la cécité verbale [The different types of aphasia—Verbal blindness]. *Progrès Médicale, 11,* 441–444.

Charcot, J. M. (1883b). Un cas de suppression brusque et isolée de la vision mentale des signes et des objets (formes et couleurs) [A case of sudden and isolated suppression of the mental vision of signs and objects (shapes and colors)]. *Progrès Médicale, 11,* 568–571.

Coltheart, M., Patterson, K. E., & Marshall, J. (1980). *Deep dyslexia.* London: Routledge & Kegan Paul.

Coslett, H. B., & Saffran, E. M. (1989). Preserved object recognition and reading comprehension in optic aphasia. *Brain, 112,* 1091–1110.

Damasio, A. R., McKee, J., & Damasio, H. (1979). Determinants of performance in color anomia. *Brain and Language, 7,* 74–85.

Davidoff, J. B., & De Bleser, R. (1993). Optic aphasia: A review of past studies and reappraisal. *Aphasiology, 7,* 135–154.

Davidoff, J. B., & De Bleser, R. (1994). Impaired picture recognition with preserved object naming and reading. *Brain and Cognition, 24,* 1–23.

Davidoff, J. B., & Donnelly, N. (1990). Object superiority effects: Complete versus part probes. *Acta Psycholinguistica, 73,* 225–243.

Denes, G., & Semenza, C. (1975). Auditory modality-specific anomia: Evidence from a case of pure word deafness. *Cortex, 11,* 401–411.

De Renzi, E. (1986). Slowly progressive visual agnosia or apraxia without dementia. *Cortex, 22,* 171–180.

Ellis, A. W. (1984). *Reading, writing and dyslexia: A cognitive analysis.* London: Lawrence Erlbaum.

Farah, M. J. (1990). *Visual agnosia: Disorders of object recognition and what they tell us about normal vision.* Cambridge, MA: MIT Press.

Farah, M. J. (1994). Neuropsychological inference with an interactive brain: A critique of the "locality" assumption. *Behavioral and Brain Sciences, 17,* 43–104.

Farah, M. J., & McClelland, J. L. (1991). A computational model of semantic memory impairment: Modality-specificity and emergent category-specificity. *Journal of Experimental Psychology: General, 120,* 339–357.

Freud, S. (1935). *On aphasia.* E. Stengel, Trans. London: Imago. (Original work published 1891)

Freund, D. C. (1888). Einige Grenzfälle zwischen Aphasie und Seelenblindheit [Some marginal cases between aphasia and mind blindness]. *Allgemeine Zeitschrift für Psychiatrie, 44,* 660–663.

Freund, D. C. (1889). Über optische Aphasie und Seelenblindheit [On optic aphasia and mind blindness]. *Archiv für Psychiatrie und Nervenkrankheiten, 20,* 276–297, 371–416.

Gil, R., Pluchon, C., Toullat, G., Michenau, D., Rogez, R., Levevre, J. P. (1985). Disconnexion visuo-verbale (aphasie optique) pour les objets, les images, les couleurs et les visages avec alexie "abstractif," [Visuo-verbal disconnection (optic aphasia) for objects, images, colors, and faces with abstract alexia]. *Neuropsychologia, 23,* 333–349.

Goodglass, H., Barton, M. I., & Kaplan, E. F. (1968). Sensory modality and object naming in aphasia. *Journal of Speech and Hearing Research, 11,* 488–496.

Goodman, R. A., & Caramazza, A. (1986). Phonologically plausible errors: Implications for a model of the phoneme–grapheme conversion mechanism in the spelling process. In G. Augst (Ed.), *New trends in graphemics and orthography* (pp. 263–296). Berlin: De Gruyter.

Grailet, J. M., Seron, X., Bruyer, R., Coyette, F., & Frederix, M. (1990). Case report of a visual integrative agnosia. *Cognitive Neuropsychology, 7,* 275–309.

Hart, J., & Gordon, B. (1992). Neural subsystems for object knowledge. *Nature, 359,* 60–64.

Hécaen, H., Goldblum, M. C., Masure, M. C., & Ramier, A. M. (1974). Une nouvelle observation d'agnosie d'objet. Déficit de l'association ou de la categorisation, spécifique de la modalité visuelle? [A new observation in object aphasia. A deficit of association or categorization specific to the visual modality?]. *Neuropsychologia, 12,* 447–464.

Hier, D., & Mohr, J. P. (1977). Incongruous oral and written naming. Evidence for a subdivision of the syndrome of Wernicke's aphasia. *Brain and Language, 4,* 115–126.

Hillis, A. E., Rapp, B., Romani, C., & Caramazza, A. (1990). Selective impairment of semantics in lexical processing. *Cognitive Neuropsychology, 7,* 191–243.

Humphreys, G. W., Riddoch, M. J., & Quinlan, P. T. (1988). Cascade processes in picture identification. *Cognitive Neuropsychology, 5,* 67–103.

Iorio, L., Falanga, A., Fragrassi, N. A., & Grossi, D. (1992). Visual associative agnosia and optic aphasia: A single case study and a review of the syndromes. *Cortex, 28,* 23–37.

Job, R., & Sartori, G. (Eds.). (1988). The cognitive neuropsychology of visual and semantic processing of concepts. *Cognitive Neuropsychology, 5,* 1–150.

Kawahata, N., & Nagata, K. (1989). A case of associative visual agnosia: Neuropsychological findings and theoretical considerations. *Journal of Clinical and Experimental Neuropsychology, 11,* 645–664.

Kay, J., & Ellis, A. (1987). A cognitive neuropsychological case study of anomia: Implications for psychological models of word retrieval. *Brain, 110,* 613–629.

Kertesz, A. (1979). Visual agnosia: The dual deficit of perception and recognition. *Cortex, 15,* 403–419.

Kleist, K. (1914). Aphasie und Geisteskrankheit. *Münchener medizinische Wochenschrift, 61,* 8–12.

Kleist, K. (1934). Kriegsverletzungen des Gehirns in ihrer Bedeutung für die Hirnlokalisation und Hirnpathologie. In B. von Schjerning & K. Bonhoeffer (Eds.), *Handbuch der ärztlichen Erfahrungen im Weltkrieg* (pp. 1–139). Leipzig: Barth.

Kremin, H. (1986). Spared naming without comprehension. *Journal of Neurolinguistics, 2,* 131–150.

Kussmaul, A. (1877). *Die Störungen der Sprache* [The disorders of language]. Leipzig: Vogel.

Lange, J. (1936). Agnosien und Apraxien. In O. Bumke & O. Foerster (Eds.), *Handbuch der Neurologie, vol. VI.* Berlin: Springer.

Larrabee, G. J., Levin, H. S., Huff, F. J., Kay, M. C., & Guinto, F. C. (1985). Visual agnosia contrasted with visual disconnection. *Neuropsychologia, 23,* 1–12.

Levine, D. N. (1978). Prosopagnosia and visual object agnosia: A behavioral study. *Brain and Language, 5,* 341–365.

Lhermitte, F., & Beauvois, M. F. (1973). A visual speech disconnexion syndrome: Report of a case with optic aphasia, agnosic alexia and colour agnosia. *Brain, 96,* 675–714.

Lhermitte, F., Chedru, F., & Chain, F. (1973). A propos d'un cas d'agnosie visuelle. *Revue Neurologique, 128,* 301–322.

Lindeboom, J., & Savinkels, J. A. (1986). Interhemispheric communication in a case of total visuo-verbal disconnection. *Neuropsychologia, 24,* 781–792.

Lissauer, H. (1890). Ein Fall von Seelenblindheit nebst einem Beitrage zur Theorie derselben [A case of mind blindness and a theoretical contribution]. *Archiv für Psychiatrie und Nervenkrankheiten, 21,* 220–270.

Mack, J. L., & Boller, F. (1977). Associative visual agnosia and its related deficits: The role of the minor hemisphere in assigning meaning to visual perceptions. *Neuropsychologia, 15,* 345–351.

Marr, D. (1982). *Vision: a computational investigation into the human representation and processing of visual information.* San Francisco: W. H. Freeman & Co.

Marshall, J. C. (1989). Commentary: Carving the cognitive chicken. *Aphasiology, 3,* 735–740.

Milner, A. D., Perett, D. I., Johnston, R. S., Benson, P. J., Jordan, T. R., Heeley, D. W., Bettucci, D., Mortara, F., Mutani, R., Terazzi, E., & Davidson, D. L. W. (1991). Perception and action in 'visual form agnosia'. *Brain, 114,* 405-428.

Morton, J. (1979). Word recognition. In J. Morton & J. C. Marshall (Eds.), *Psycholinguistics* (Series 2) (pp. 105–156). London: Elek.

Morton, J. (1980a). The logogen model and orthographic structure. In U. Frith (Ed.), *Cognitive processes in spelling.* London: Academic Press.

Morton, J. (1980b). Two auditory parallels to deep dyslexia. In M. Coltheart, K. E. Patterson, & J. C. Marshall (Eds.), *Deep dyslexia* (pp. 189–197). London: Routledge & Kegan Paul.

Morton, J. (1985). Naming. In S. Newman & R. Epstein (Eds.), *Current perspectives in dysphasia.* Edinburgh: Churchill Livingstone.

Morton, J., & Patterson, K. (1980). A new attempt at an interpretation, or, an attempt at a new interpretation. In M. Coltheart, K. E. Patterson, & J. C. Marshall (Eds.), *Deep dyslexia* (pp. 91–118). London: Routledge & Kegan Paul.

Newcombe, F., & Marshall, J. C. (1980). Transcoding and lexical stabilisation. In M. Coltheart, K. E. Patterson, & J. C. Marshall (Eds.), *Deep dyslexia* (pp. 176–189). London: Routledge & Kegan Paul.

Newcombe, F., & Ratcliff, G. (1974). Agnosia: A disorder of object recognition. In F. Michel & B. Schott (Eds.), *Les syndromes de disconnexion calleuse chez l'homme* Lyon: Colloque International de Lyon.

Oxbury, J. M., Oxbury, S. M., & Humphrey, N. K. (1969). Varieties of colour anomia. *Brain, 92,* 847–860.

Patterson, K. E. (1988). Acquired disorders of spelling. In G. Denes, C. Semenza, & P. Bisiachi (Eds.), *Perspectives on cognitive neuropsychology* (pp. 213–228). London: Lawrence Erlbaum.

Patterson, K. E., Marshall, J. C., & Coltheart, M. (Eds.). (1985). *Surface dyslexia: Neuropsychological and cognitive studies of phonological reading.* London: Lawrence Erlbaum.

Pena-Casanova, J., Roig-Rovira, T., Bermudez, A., & Tolosa-Sarro, E. (1985). Optic aphasia, optic apraxia, and loss of dreaming. *Brain and Language, 26,* 63–71.

Rapp, B., & Caramazza, A. (1993). On the distinction between deficits of access and deficits of storage: A question of theory. *Cognitive Neuropsychology, 10,* 113–141.

Rapp, B. C., Hillis, A. E., & Caramazza, A. (1993). The role of representations in cognitive theory: More on multiple semantics and the agnosias. *Cognitive Neuropsychology, 10,* 235–249.

Ratcliff, J., & Newcombe, F. (1982). Object recognition: Some deduction from the clinical evidence. In A. W. Ellis (Ed.), *Normality and pathology in cognitive functions* (pp. 147–171). London: Academic Press.

Riddoch, M. J., & Humphreys, G. W. (1987a). A case of integrative visual agnosia. *Brain, 110,* 1431–1462.

Riddoch, M. J., & Humphreys, G. W. (1987b). Visual object processing in optic aphasia: A case of semantic access agnosia. *Cognitive Neuropsychology, 4,* 131–185.

Riddoch, M. J., Humphreys, G. W., Coltheart, M., & Funnell, M. (1988). Semantic systems or system? Neuropsychological evidence re-examined. *Cognitive Neuropsychology, 5,* 3–25.

Saussure, F. de (1959). *Course in general linguistics* (W. Baskin, Trans.). New York: Philosophical Library. (Original work published 1916)

Seymour, P. H. K. (1979). *Human visual cognition.* London: Collier Macmillan.

Schriefers, H., Meyer, A. S., & Levelt, W. J. M. (1990). Exploring the time course of lexical access in language production: Picture–word interference studies. *Journal of Memory and Language, 29,* 86–102.

Shallice, T. (1988). *From neuropsychology to mental structure.* Cambridge, UK: Cambridge University Press.

Shallice, T. (1993). Multiple semantics: Whose confusions? *Cognitive Neuropsychology, 10,* 251–261.

Shuren, J., Geldmacher, D., & Heilman, K. M. (1993). Nonoptic aphasia: Aphasia with preserved confrontation naming in Alzheimer's disease. *Neurology, 43,* 1900–1907.

Sirigu, A., Duhamel, J-R, & Poncet, M. (1991). The role of sensorimotor experience in object recognition. *Brain, 114,* 2555–2573.

Spreen, O., Benton, A. L., & Van Allen, M. W. (1966). Dissociation of visual and tactile naming in amnesic aphasia. *Neurology, 16,* 807–814.

Taylor, A. M., & Warrington, E. K. (1971). Visual agnosia: A single case report. *Cortex, 7,* 152–161.

Warren, C. E. J., & Morton, J. (1982). The effects of priming on picture recognition. *British Journal of Psychology, 73,* 117–130.

Warrington, E. K., & James, M. (1988). Visual apperceptive agnosia: A clinico-anatomical study of three cases. *Cortex, 24,* 13–32.

Warrington, E. K., & McCarthy, R. (1983). Category-specific access dysphasia. *Brain, 106,* 859–878.

Warrington, E. K., & McCarthy, R. (1987). Categories of knowledge. *Brain, 110,* 1273–1296.

Warrington, E. K., & Shallice, T. (1984). Category-specific semantic impairments. *Brain, 107,* 829–854.

Wernicke, C. (1874). *Der aphasische Symptomencomplex. Eine psychologische Studie auf anatomischer Basis* [The aphasic symptom complex. A psychological study on an anatomical basis]. Breslau: Cohn & Weigert.

Wernicke, C. (1886). Nervenheilkunde. Die neueren Arbeiten über Aphasie. *Fortschritte der Medicin, 4,* 371–377.

Wernicke, C. (1906). Der aphasische Symptomencomplex [The aphasic symptom complex]. In E. von Leyden & F. Klemperer (Eds.), *Die deutsche Klinik am Eingange des zwanzigsten Jahrhunderts in akademischen Vorlesungen, Volume VI, Nervenkrankheiten* [The German clinic at the beginning of the twentieth century in academic lectures, Volume VI, nervous diseases] (pp. 487–556). Berlin: Urban & Schwarzenberg.

Wolff, G. (1904). *Klinische und kritische Beiträge zur Lehre von den Sprachstörungen* [Clinical and critical contributions to the study of language impairments]. Leipzig: Veit.

Chapter 5

Proper-Name-Specific Aphasias

Carlo Semenza

THE SPECIFICITY AND THE IMPORTANCE OF PROPER NAMES

The use of proper names is a necessary part of social communication in many different contexts and cultures. However, proper names often pose a difficult retrieval challenge and, more often than other nouns, make one subject to temporary failures. Attempts to circumvent this sort of problem may be painstaking or result in ambiguous identification. This vulnerability therefore affects the success of communication and may cause confusion, embarrassment, and offense. It is common knowledge that many elderly persons, even when they are quite intact in terms of basic cognitive functions, complain of and, indeed, are prone to retrieval failures of proper names. This consequence of age-related changes in memory ability adds a nontrivial difficulty to old people's social adjustment.

Psychologists have provided so far a considerable amount of laboratory and naturalistic data demonstrating the propensity for proper names to be forgotten. The reasons for this rather ubiquitous finding have been, however, neglected for a long while. The possibility that the greater difficulty would derive from a sort of processing different from that of common names was not seriously considered until rather recently. Most theoretically and empirically relevant works, indeed have appeared only in the last decade. Some remarkable findings in brain-damaged patients have put neuropsychologists on the front line of this kind of research. A characteristic of this revived interest in proper names is the convergence of the theories and, indeed, of empirical support with those of philosophers that, for over a century and a half, never ceased to consider proper names as theoretically important. Modern linguistics is also presently contributing to the issue of proper names in an important way.

This chapter will report proper-name-related phenomena in aphasia as they are now known, with a special eye to the theoretical implications of such findings in the more general philosophical and linguistic context. The point will be made that an understanding of the differences in processing

*ANOMIA: Neuroanatomical
and Cognitive Correlates*

115

between common and proper names allows an unprecedented level of insight about the working of the semantic system in general. Proper names are important not only for social reasons or because they are a special case in semantics, but because by comparing their processing with that of common names, a better view will be obtained of the name-retrieval process.

THE THEORETICAL BACKGROUND

In 1843 Mill wrote that "proper names are not connotative: they denote individuals who are called by them: but they do not indicate or imply any attributes as belonging to those individuals. . . . A proper name is but an unmeaning mark." Frege's (1892) later distinction between two aspects of meaning, "sense" and "reference," helped to clarify the issue, and was adopted by later philosophers like Wittgenstein (1922) and, in contemporary times, Kripke (1980). They argued that proper names carry "reference," that is, they denote the individuals or the entities that are called by them, but have no "sense," insofar as they do not describe any property or imply any attribute. For some authors (like Kripke, 1980) this makes proper names just the opposite of "descriptions," which have sense and encompass all common nouns. The alternative view, held for example by Russel (1905) and, in more recent times, by Searle (1969), is that proper names are also descriptions, albeit very short ones, carrying only a minimum of sense. For the purposes of the present review these alternatives can be safely ignored. It will suffice to regard proper names as bearing little if any meaning at all; the link with their referent being weaker and more arbitrary than that between a common name and its referent.

The above distinctions between proper and common names are better illustrated by the following examples. The proper name "Bill Clinton" refers only to the bearer of this name and does not itself provide any other information. On the other hand, the word *president* is a description: it implies the definition of a person who is chief of a nation or a club, has more or less specific powers, and so on. The limited truth value that proper names have may be exemplified by the fact that anyone would understand fully sentences of the type, "There are no popes in Australia," whereas they would not fully understand sentences like, "There are no Wojtylas in Australia," unless they knew that Wojtyla is actually the name of the pope. For the same reasons, changing proper names does not alter, as with common names, any property of the bearer. A woman that drops her maiden name for that of her husband (there are still places, like Austria, where this is more or less compulsory or, however, overwhelmingly in custom) does not become a different person by this very act, although she acquires certain properties and loses others becoming a "wife." Thus "Karol Wojtyla" did not become a different person by changing his name to "John Paul II," but acquired a very different status when, being formerly a cardinal, he was made pope. On the other hand, proper names refer to a unique bearer indifferently across all the actual and counterfactual situations, past, present, and future of which the bearer is a constituent. A biographer of Karol Wojtyla would therefore use the proper name Wojtyla for the boy that grew up in Poland, the young priest, the cardinal in Crakow, and the chief of the Catholic Church in Rome in a given period. In the same way it is perfectly proper to say that John Paul II was born before Second World War, although he was not called John Paul II until 1978.

Modern (psycho)linguistics makes similar arguments. Thus Miller and Johnson-Laird (1976) argued that proper names have a high probability of having a unique referent. An important distinction in conceptual structure is the binary feature type or token (Jackendoff, 1983; Katz, 1972; Levelt, 1989). What one learns and stores in memory can be linked either with token (if one is remembering an individual) or with type (if one has learned a category). Proper names, because they denote individuals in a category (identified as a basic object level kind by MacNamara, 1982, and La Palme-Reyes, MacNamara, Reyes, & Zolfaghari, 1993), have only token reference and not type reference.

THE EXPERIMENTAL LITERATURE

The experimental literature on proper names is vast and included here are only the most important and recent findings (the reader is referred to Cohen & Burke, 1993, and Valentine, Brennen, & Brédart, 1996, for a more comprehensive review). Some research has been concerned with demonstrating that naturally occurring retrieval blocks are much more frequent for proper names than for other kinds of words (Bolla, Lindgren, Bonaccorsi, & Blecker, 1991; Reason & Lucas, 1984; Young, Hay, & Ellis, 1985). Cohen and Faulkner (1986) confirmed this pattern in a more formal experimental study.

Proper names have also been demonstrated to be vulnerable to the tip-of-the-tongue (TOT) phenomenon by Burke, MacKay, Worthley, and Wade (1991). These authors provided an explanation for this finding in terms of an interactive activation theory of language production known as the Node Structure Theory (NST) (MacKay, 1987). According to NST the activation of a lexical node for a common name (e.g., baker) would benefit by several converging semantic connections from the semantic system, thus being relatively invulnerable to TOT. The activation of a lexical node representing a proper name (e.g., Baker) is thought, instead, to spread from the semantic system to the lexical node only via propositional nodes for specific individuals ("John Baker," "Mary Baker," etc.). These latter nodes may, indeed, receive even abundant converging semantic information about the individual, but there would be a single, and therefore vulnerable, connection in their output for the activation of the phonological form in the corresponding lexical node. Thus, even though the bearer of the name is highly familiar, his or her name (e.g., Baker) is more prone to TOT than, for instance, his or her occupation (e.g., baker). This is, indeed, an explanation that also applies to the so-called Baker–baker paradox found in tasks that require learning names and occupations belonging to unfamiliar faces. As repeatedly demonstrated (Cohen, 1990; McWeeny, Young, Hay, & Ellis, 1986) a word like *Baker* presented as a name is, in fact, harder to recall than the same word (*baker*) presented as an occupation. At this point it is important to note (see also Cohen, 1990) that such findings are perfectly compatible with the philosophers' idea described above that the link a proper name entertains with its referent is a weak and arbitrary one.

According to most researchers (Cohen, 1994), elderly people experience an increased difficulty in recalling proper names. However, few systematic studies have been devoted to determining whether this widespread belief is in fact well grounded. Indeed, younger subjects perform better than older subjects (Crook & West, 1990), but no data were available until very recently on whether any decline in performance due to age differs between proper and common names. The main problem is to make sure that the naming tasks used to compare common and proper-name retrieval are of comparable difficulty. It is unclear, for instance, how this could be obtained with pictures of faces on the one side and pictures of objects on the other. The solution to this problem was sought by Semenza, Nichelli, and Gamboz (1996) by adopting a free recall test of supraspan lists of names. In these lists, common and proper names were directly matched for frequency, length, and phonological complexity. A better primacy effect was found for common names at all ages. However, this difference became larger in subjects older than 70.

Other investigators are concerned with proper names, mainly outlining the process of generating the proper name corresponding to a face. A model has been proposed (Bruce & Young, 1986) consisting of a sequence of functional components that is common to the recognition and naming of objects and words. The sequence comprises formation of an input code; activation of a face-recognition unit; access to semantic information including a person's biographical and contextual information; and, finally, access to the person's name. This final stage in the sequence can only be performed via semantic information, and there is no direct link between faces and names. The model accounts for the fact that, although it is often the case that one does not seem to remember a name, but can remember biographical details about a person (a phenomenon shown in certain types of anomias—e.g., by Flude, Ellis, & Kay, 1988,—and dramatically evident in selective proper name anomias), the converse type of the incident, in which the name is known but biographical details cannot be recalled, appears to be extremely rare (Young et al., 1985). What is particularly interesting in this model is the idea of specific units mainly devoted to the storage of identity-specific semantic information. In the most articulate version of the same model (due Valentine, Brédart, Lawson, & Ward, 1991), those words in the input lexicon that are not proper names pass activation to a pool of units called "word-specific semantics." Words that are proper names, instead, pass activation to a new set of units, name-recognition units, which can feed activation to identity-specific semantic information units. An interesting feature of this latter version is that the lexical output is thus activated separately from proper-name-recognition units and word-recognition units.

In a series of works, Burton and Bruce (e.g., 1992, 1993) argued, however, that a hierarchically organized architecture, even if consistent with empirical data, may not be entirely adequate. They pointed out that we do not have to retrieve *everything* about a person before retrieving their name. A simple linear process by which *all* personal information is accessed before a name is inadequate. They consequently built up an interactive activation and competition model, that keeps a distinction among different pools of word-recognition units, (proper) name recognition units, and personal identity units. In each pool each unit entertains a bidirectional connection with the corresponding elements in adjacent pools. Personal identity nodes are still connected to word-recognition units only via name-recognition units. With such a model they could simulate a series of empirical findings like the advantage in retrieval of known over unknown and of common over uncommon names. They were also able to test whether or not name-recognition units were necessary in the model: the results of simulation confirmed the necessity of such units.

NEUROPSYCHOLOGICAL FINDINGS

With proper-name-specific aphasias we refer to a set of aphasic phenomena whereby proper names are, relative to common names, either selectively impaired or spared in linguistic output or input. With only one known exception, these phenomena have been reported very recently and appear to be somehow rare. However, since the appearance of the first extensive study of proper-name anomia (Semenza & Zettin, 1988), a number of aphasias concerning the productive side have been reported in the literature[1] with increased frequency. It is thus possible that these phenomena are not so rare, but that for a long time they passed unobserved because

---

[1] I have now also accumulated an almost equal number of anecdotal yet reliable case reports that are not yet published.

proper names are not systematically studied and clinical batteries put little, if any, emphasis on them. As indeed happened, cases had to appear, in order to be noticed, where the dissociation between proper names and common names was virtually absolute and constituting the only serious symptom and the only concern of the patient.

Proper name-related neuropsychological phenomena are not, however, just anomias. Previous to recent, more revealing case reports, evidence had been provided (Saffran, Bogyo, Schwartz, & Marin, 1980) that deep dyslexics read proper names surprisingly well. This observation converged with the finding that normal readers read proper names very well in the left visual field. These results were interpreted as indicating superior reading of proper names compared with common ones by the right hemisphere. Other evidence of superiority of proper names over common ones was later reported by Van Lanker and Klein (1990) in the task of matching spoken or written names to photographs in four cases of global aphasia. These authors pointed out that qualities like familiarity or affection would allow some advantage for proper names in right-hemisphere processing. A relative preservation for a particular category of proper names, those of places, was also shown in some aphasics in the task of matching them to a map (Wapner & Gardner, 1979). Spoken-to-written matching was also shown to be surprisingly well preserved in a very severe global aphasia for certain categories of proper names (those of famous people, countries, and cities) by Warrington and McCarthy (1983) and more recently, in another severely impaired case described by McNeil, Cipolotti, and Warrington (1993). Finally Goodglass and Wingfield (1993) showed an interesting dissociation between the comprehension of geographical names selectively preserved in Wernicke's aphasics, and body part names selectively preserved in anomics.

It thus seems that for tasks like matching, where the retrieval of the phonological form is not requested, or reading, where it can be derived from the orthographic form, patient studies converge in indicating a superiority in performance involving proper names in comparison to common ones. In naming tasks, however, where the phonological form is not provided and cannot be more or less derived from the orthographical one, the story is different and, to this day, far more interesting.

As it has been said, not all of the data on anomias are new. In 1683, in fact J. J. Wepfer, a neurologist in Schauffausen,[2] observed the case of R. N. N., a 53-year-old man, who in the course of the recovery from what was probably an acute brain bleed, improved from an initially very severe aphasia to a more selective and unusual disorder. In fact, although his au-

---

[2]I owe this information to Claudio Luzzatti, who discovered Wepfer's works in the Brera Museum in Milan, Italy. The interested reader is referred to Luzzatti and Whitaker's (1996) comprehensive report of Wepfer's studies.

ditory comprehension, repetition, reading, writing, and object naming recovered quite nicely, and he was left with what appears to have been a certain degree of agrammatism, he could not find proper names of humans and places as quickly as usually. Details are not provided, but this case may probably be considered the first report of anomia where proper names were disproportionately more affected than common ones (observation 98, posthumously published in Wepfer, 1727).

In 1980 McKenna and Warrington described the first case where only the names of persons were disturbed. The deficit appeared to be equally severe whether naming was from a picture or from a verbal description. The comprehension of the same names and knowledge about the target people was shown to be perfectly preserved. The first modern cases where both persons' and geographical names were selectively disturbed were reported in two very detailed studies by Semenza and Zettin (1988, 1989). These cases were peculiar in that, although the anomia for proper names was extremely severe, common names, even very difficult, abstract ones, were fairly well preserved. The anomia, which spared just a few items of autobiographical value (little more than the patients' own names, those of some close relatives, and of their hometowns), was present in both oral and written modality over a series of conditions, including confrontation naming and naming on description and by category. Conversely, reading aloud (e.g., naming of written words) was also perfect for irregularly spelled names. In some cases phonological forms were presented in both a proper name and in a common name definition context. The patients (this was later replicated more formally by Hittmair-Delazer, Denes, Semenza, & Mantovani, 1994) could only retrieve the name when it was in a common-name context. For example, they could answer *"colombo"* (pigeon) to the question, "What is the sort of bird living in San Marco Square in Venice?" but could not retrieve "Colombo" (Columbus) when asked "Who discovered America?" No other clinical disturbance was reported by the patients, nor could it be detected via extensive traditional testing. Contemporary theoretical speculations and theory-driven empirical research on the topic of proper names in neuropsychology (see later in the chapter) started with these two studies, which were later followed by a number of other case descriptions where the defect was more evident for persons (the first clear-cut case being that reported by Lucchelli & De Renzi, 1992) and much less for the names of places, which appeared relatively preserved. In one case (Carney & Temple, 1993), this pattern of "person-only" anomia was interpreted as being specific for faces and called a case of "prosopanomia." In order to support this interpretation, however, the authors should have shown that the patient was perfectly able to provide proper names from verbal description. Unfortunately they did not conduct this experiment, and, indeed, the scanty available evidence seems to show that the patient is similar to Lucchelli and De Renzi's (1992) and other cases of persons-only anomia.

All these patients seemed to be able to repeat names, but they would forget them after any small amount of delay: in some patients 15 sec of counting backward from 100 was enough to forget the name they had just pronounced and repeated to themselves a few times. This was not the case with others words. Hittmair-Delazer et al. (1994) also showed how proper-name anomic M. P., in repeating supraspan lists of words, lacked a primacy effect if lists consisted of proper names.

Another interesting finding with these patients is their sensitivity to cuing. Phonemic cuing alone was not effective at all in the severest patients (Semenza & Zettin, 1988, 1989), but in others (e.g., in Cohen, Bolgert, Timsit, & Cherman, 1994) it dramatically reduced the degree of anomia. Even the severest patients, however, were sensitive to semantic cuing, but only when the proper names also had a real meaning. Thus, Semenza and Zettin's patients improved their performance to about 25% correct for names like "Colombo" on questions like, "Tell me who discovered America; he had a name of a bird." The best performance (around 50%) was however achieved by these patients when both this sort of semantic cuing and phonemic cuing were provided. Hittmair-Delazer et al. (1994) also cued their patient M. P. with first names, thus obtaining an increment in performance from 0 to 100% in a confrontation naming task. This was held to indicate that representations labeling individuals are a unit prior to phonological activation and comprise first and second name. A similar explanation may be given to the phenomenon described by Damasio and Tranel (1990) in their patient Boswell. This patient could easily complete, once given the city element, city–state pairings of the sort Denver–Colorado, while demonstrating no knowledge of either the city or state. Knowledge of the first–second names or of the city–state name pairings is thus probably acquired and stored in this particular order as a single representational unit that is independent of any other type of knowledge.

A summary description of all published cases (updated to 1995) of genuine anomias for proper names is reported in Table 1. A case that does not fit the general pattern but that, with shallow scrutiny, could be misunderstood for a typical example of anomia for proper names, was reported by Shallice and Kartsounis (1993). Indeed their patient suffered from an anterograde retrieval disorder for names acquired in the past 20 years. As Shallice and Kartsounis demonstrated, this disturbance concerned a large number of names of people that became famous after the onset of the patient's disease (hence an early suspicion of a proper-name anomia) but also newly emerged elements of the vocabulary like "AIDS." This was not, obviously, the case with authentic proper-name anomias, where retrieval difficulties concerned lifelong known proper names.

The reverse pattern, anomia for common names relative to proper names (see Table 2) appears to be rare. McKenna and Warrington (1978) reported

Table 1 Proper Names Deficit

	Proper names category	Associated problems	Lesion
Wepfer (1727)	Persons and geographic names	Unknown	Unknown
McKenna & Warrington (1980)	Persons	Unknown	Region of the posterior temporal branch of the middle cerebral artery
Semenza & Zettin (1988)	Persons and geographic names	Unknown	Left parieto-occipital
Semenza & Zettin (1989)	Persons and geographic names	Inability to learn arbitrary links between word Inability to tell titles of known pieces of music Inability to learn number labels to various items	Left fronto-temporal (with involvement of basal structures)
Lucchelli & De Renzi (1992)	Persons	Inability to learn name–face and number–color pairs Inability to recall previously known telephone numbers	Left thalamus
Carney & Temple (1993)	Persons	Unknown	Multiple
Hittmair-Delazer, Denes, Semenza, & Mantovani (1994)	Persons	Inability to learn arbitrary links between words Inability to learn name–face pairs Inability to retrieve personal number facts	Left fronto-temporal, including basal ganglia
Cohen, Bolgert, Timsit, & Cherman (1994)	Persons	Unknown	Left thalamus (affecting VA, VL nuclei) and mammillo-thalamic tract, possibly part of DM nucleus
Fery, Vincent, & Bredart (1995)	Persons	Poor learning of associate word pairs	Mild left cerebral athrophy and genu of internal capsule
Harris & Kay (1995)	Persons and geographic names	Inability to learn verbal associations	Left temporal

the case of a patient whose naming of body parts and of a limited number of high-frequency names of countries was much superior to naming of various other categories. Warrington and Clegg (1993) reported another case of preservation for the names of countries, which were named significantly better than colors, objects, animals, and body parts. Two other cases concerned persons (in Semenza & Sgaramella, 1993, places were untestable due to the patient's bad knowledge of geography) and persons and countries as well (Cipolotti, McNeil, & Warrington, 1993). In both these cases the patients' output was extremely disturbed. R. I., Semenza and Sgaramella's patient, could spontaneously produce only meaningless monosyllables, frequently intercalated with person names. His agraphia was such that he could not even reproduce single letters; thus his written naming performance could not be tested. Confrontation naming as well as naming from definition for both common and proper names appeared hopelessly substituted with randomly produced monosyllables. In an attempt at understanding how the correct phonological form of proper names appeared in spontaneous output, Semenza and Sgaramella tested naming aided by phonemic cuing. The situation changed dramatically. When provided with the first phoneme followed by a schwa, R. I. correctly produced, on both confrontation and definition the names of all the people he knew, but he could not produce the name of even the commonest objects, resorting in this case to his monosyllabic jargon. Cipolotti et al.'s (1993) patient M. E. D. was also atypical. Although her spoken-name retrieval was severely compromised whatever the type of input material, her ability to write the names of countries and famous people was consistently superior to her ability to write the names of objects.

In conclusion, cases of proper-name preservations are, as already told, rarer and, so far, less clear than anomias. Indeed, although in two cases sparing was demonstrated for a limited number of high-frequency names of countries, in the other two cases sparing was shown to concern most persons—

Table 2   Proper Names Sparing

	Proper names category	Lesion
McKenna & Warrington (1978)	Countries	Left temporal
Semenza & Sgaramella (1993)	Persons	Parieto-occipital
Cipolotti, McNeil, & Warrington (1993)	Persons, countries	Left fronto-parietal and thalamus
Warrington & Clegg (1993)	Countries	Extended cortical atrophy, most prominently left temporal

acquaintances and famous people—known to the subject. There is a confusing element in these last two cases, in the specific output conditions where sparing was exclusively shown: phonemically cued output in R. I. and written output in M. E. D. Uncertainty about the precise locus of impairment in these cases (both sets of authors have specific theories about what happened in their patients—see also later in this chapter) cannot, however, invalidate the fact that, taken together with anomias for proper names, they constitute a theoretically important double dissociation in lexical output.

ANATOMICAL LESIONS AND THE LOCALIZATION OF FUNCTIONS

The observed double dissociation between common and proper names may indicate that the nervous system honors the distinction between these two categories by processing them in separate structures. Can one go further and identify precisely which different part subserves which category of names? A similar question has been recently positively answered for the distinction between object or entity names and the names of actions (i.e., verbs). Thus Damasio and Tranel (1993) claimed that the left anterior and middle temporal lobe contains a system for the retrieval of nouns that denote concrete entities; the equivalent mediation systems for verbs would instead be located in the left frontal region. The merit of this claim will not be discussed here, where the issue addressed is of whether it is possible to identify in similar detail the different regions where proper names and common names are stored and processed. Indeed Damasio et al. produce some evidence indicating that the left temporal polar cortex is important for the retrieval of proper names but not for the retrieval of common names. Their data came from two parallel studies, one conducted in brain-damaged patients and the other using PET scans in normal individuals. Out of 127 patients, 7 subjects were identified who performed significantly worse in proper names than on the other names in a visual confrontation naming task. All these subjects had a lesion in the pole of the left temporal lobe, which also happened to be the region that was most activated in a person-naming task by normals. Of course, the left temporal pole might be involved in the system of proper-name retrieval, but the story is certainly more complex, as all reported cases of proper-name-specific aphasias will tell.

Table 1 shows all known cases of anomia selective for proper names, where recognition of the same items was unimpaired. These cases are more or less pure, in the sense that the patients had virtually no other aphasic symptom whatsoever. Despite this striking similarity among patients, their lesions are quite scattered in the left hemisphere. Their location seems to be somehow marginal with respect to the perisylvian areas, and their bulk is always outside the frontal lobe. Most but not all cases (which may be due to inadequate imaging) appear to involve the temporal lobe. In a good proportion of the cases there is sign of involvement of deep structures, particularly of the thalamus. If one then considers the few cases of selective sparing (Table 2), the topography of the corresponding lesions does not seem to be more revealing: indeed, again, the foci of the lesions are scattered around in the same areas where damage has been demonstrated to lead to the opposite phenomenon.

Vis-à-vis this sort of evidence, any attempt at locating more precisely the sort of structure responsible for proper-names retrieval or for the retrieval of terms with pure referential value seems premature. Yet, because it appears to be quite a distinct structure (one would be tempted to call it a "module," if agreement could be found about what a module is—see Semenza, 1996), the effort of locating it in the brain must not be given up. Two theoretical alternatives seem, however, to be the more likely ones. The first would be a structure constituted by a rather compact set of neurons, that, subject to individual variation, is located in areas that are marginal to the main perisylvian language areas. Over about this same area, according to the second alternative, the critical components would be located in a distributed neural network: in this case, selective damage to different portions of the network may cause the same epiphenomenon. Further detail will be needed to accommodate for distinctions like the ability to retrieve either all proper names of person or geographical names only. Also the sensitivity to cuing may play a role in variation, as proposed by Cohen et al. (1994). They observed that in cases where a thalamic lesion was found, phonemic cuing was very effective. Accordingly, they argued that phonological cuing may compensate for the lack of a sufficient activation of the output processes that result from a thalamic lesion.

THE ARBITRARY LINK HYPOTHESIS

It has been argued in the preceding section that a dedicated system seems to exist somewhere in the brain for the retrieval of proper names. Does this system work only for this specific, evolutionarily very sophisticated ability or is proper-name retrieval just one aspect of a more generic primitive function? Philosophers have claimed, as previously described, that the link of a proper name with its reference is "arbitrary." This means essentially that proper names indicate single individuals and not categories, the link being thus one-to-one without any description allowing generalization to other items. A system dedicated to proper-name retrieval could thus be described as a system able to retrieve item information in a one-to-one fashion. The case of proper names and their references is not, in this respect, unique.

Consider a task like the paired associate learning from the Wechsler Memory Scale or similar tasks requiring the retrieval of one component of a word pair. If the two components are not semantically related, the link between them is arbitrary. Semenza and Zettin (1989) first administered such a task to their patient L. S., who displayed an extremely severe selective anomia for proper names but was singularly free from other linguistic and cognitive disorders. Failures in retrieving the arbitrary pairs was nevertheless found: L. S. could learn and remember all words in the test but he was not able, over several attempts, to retrieve the second member of the pair, given the first. Converging evidence for the importance of this finding came also when the patient went back to work. He found great difficulty in learning the labeled numbers needed for hardware stock identification. L. S. was also familiar with classical music. Semenza and Zettin could thus confront him with several wordless pieces, some of which he demonstrated to know very well. Again, as expected given the perfectly arbitrary link a title has with a piece of music, L. S. was unable to retrieve any title. He could, however, recognize what he missed in naming in a multiple-choice setting.

On the basis of these observations, Semenza and Zettin (1989) tentatively attributed their patient's problem to a difficulty in retrieving components in arbitrary links. Other evidence favoring this hypothesis clearly emerged in the latter cases (see Table 1). Thus Lucchelli and De Renzi (1992) could show, in their case, an inability to learn name–face and number–color pairs and an inability to recall previously known telephone numbers. Hittmair-Delazer et al.'s (1994) patient was also disturbed in paired associate learning, could not retrieve personal number facts, and had a deficit in matching faces to names and to occupations. Similar findings are reported in Fery, Vincent, and Brédart (1995) and Harris and Kay (1995).

In conclusion, the presence of these (and not other) symptoms along with anomia for proper names can hardly be considered as merely coincidental. Nor can it be safely claimed that the anatomical region subserving both name retrieval and the retrieval of simple components of arbitrary semantic links is the same by chance. As shown above, anomia for proper names follows lesions in disparate regions albeit mostly around the temporal lobe. Chance would not allow for a systematic covariation of two independent symptoms. It is more parsimonious, and consistent with the theory, to propose that the two symptoms are two aspects of the same problem: a problem following damage to a unique processing device.

LOCATING THE FUNCTIONAL LESIONS IN PROPER-NAME ANOMIA AND AN ACCOUNT OF PROPER-NAME PRODUCTION

Separate production processing for proper and common names is suggested by the observed double dissociation. We have also seen that where and how this separation occurs in the brain is still difficult to establish. What about, then, the functional separation within a cognitivist information-processing model? Available data are, indeed, quite informative. Before, however, providing a full theoretical account, it is important to observe that a double dissociation was hardly needed to conclude exclusively on the basis of neuropsychological data that common and proper names are separately processed. The observations on proper-name anomias were enough. This is counterintuitive: one might have argued, with abundant experimental support, that proper names are more difficult to retrieve. Accordingly, a mild anomia, affecting only difficult items, would produce a proper-name anomia affecting only difficult items. This is very probably the main cause for the frequent temporary failures with proper names in the elderly (and in the non-elderly as well!). Indeed, on the basis of this consideration, proper-name anomias were not sufficient to indicate separate processing for proper names. A compelling reason, however, to reach a positive conclusion even in the absence of observations of the opposite phenomenon—anomia for common names only—came from a remarkable feature of the first-described proper-name anomias: their absolute purity. As already pointed out, virtually no proper name (at least of persons) could be retrieved by patients who succeeded in 100% of the cases with common ones. A pattern where one's mother's name is forgotten and the name of abstract, uncommon words can be easily retrieved cannot be explained with different degrees of difficulty. Separate production mechanisms for proper and common names seemed thus the only possible explanation even for the earlier neuropsychological findings.

Where then, and, how does this separation take place in the course of processing? The account provided by Semenza and Sgaramella (1993) and by Hittmair-Delazer et al. (1994) will be essentially followed here. In proper-name anomia the deficit seems to affect the retrieval of the phonological form from an intact semantic memory. Semantic information is still present and well organized as demonstrated, by a good comprehension of the auditory form and by the good amount of knowledge the patients display about the items they cannot name. The deficit, nonetheless, seems to occur prior to the input into the phonological and orthographic lexicons because the anomia follows exactly the same pattern in both the oral and the written modalities. The patients could also easily read aloud irregularly spelt proper names, thus indicating that their phonological output lexicon contains correct information. Locating the deficit at the output from the semantic system leads necessarily to the conclusion, vis-à-vis observed dissociations, that common and proper names access the lexical level from the semantic system independently from one another. Indeed, processes taking place between the semantic system and the lexicon(s) are left, with few exceptions, underspecified in models of naming currently used in neuropsychology. The simpler mechanism that has been suggested to work at this level (Butterworth, 1989) matches semantic information with the appropriate entry in the lexicon. To account for data brought about by anomias for proper names (and their selective sparing) one needs to postulate two types of such a mechanism: one for common names and one for proper names. As it will later be argued, the reasons why a given name is processed by one or other of these two mechanisms is determined by the organization of semantic memory, which may be particular for information concerning individuals. This would vary especially in the output according to the amount of descriptive information or "sense" as opposed to "reference." In the case of proper name anomias, the phonological lexicon can be activated, for proper names, only via repetition or reading and not from long-term memory. Once in the long-term store, persons' names are for such patients hopelessly impossible to retrieve (as shown by the lack of primacy effect in reporting supraspan lists of proper names demonstrated in Hittmair-Delazer et al.'s 1994 patient).

An analogous interpretation can be offered on the basis of Levelt's (1989) model of speech production, whereby lexical access is also split into two stages. The first stage consists of the retrieval of a "lemma," which is an abstract lexical item supplied with both syntactic and semantic features. The second stage of lexical access consists of access to the morphophonological form, the "lexeme." Each lemma points to its corresponding form (i.e., it can refer to the address in the form lexicon [the lexeme] where the information for that stage would include tagging of the syntactic class of a lexical item). The class of nouns is indeed divided into two major classes: proper nouns and common nouns. (This distinction is also based on the consideration that the two classes have different syntactic requirements: for instance, in English, proper names do not take the article). It is possible, then, that in proper-name anomias, processing would be selectively damaged within this stage for the class of proper names only, maybe in pointing to the lexeme form.

The damage in the case of patients with selective sparing of proper names appears to be more difficult to locate. In R. I., Semenza and Sgaramella's (1993) patient, the authors tentatively located the deficit after the lexical level prior to the realization of the articulatory program. The benefit from cuing and repetition could work indeed only at this point. Why then were proper names spared? Semenza and Sgaramella provided two possible explanations. According to the first (which is substantially similar to that of Cipolotti et al., 1993, for their case of proper-name sparing), there was additional disturbance in R. I. for common-name processing (meaning that separate processing percolates from the semantic output throughout all output levels). According to the second explanation, the proper-name process would be intrinsically more efficient if helped by phonemic cuing. Indeed, unpublished experiments conducted in their laboratory show that naming reaction times in response to a phonemic cue were faster for proper names than for common names matched for familiarity.

One last problem concerns why geographical items are often spared. As Lucchelli and De Renzi (1992) argued, this may be due to a severity effect. Why would geographical items be easier? An important factor may be the possibility of their being adjectivized. Indeed person names can also be adjectivized, but only in rare instances (e.g., "Freudian"). We have speculated that in anomia for proper names the deficit consists of a block for the grammatical class of proper names in the lexeme activation from the lemma level. It is intriguing to go further in speculating that those patients that do not have problems with geographical items overcome the retrieval block with the help of the corresponding adjective. If this is true, geographical names (e.g., those of mountains) that are not adjectivized would still be unlikely to be retrieved. This hypothesis has never been explored, and it may not be easy to test because patients with excellent geographical knowledge would be necessary.

MORE COMPLEX PHENOMENA AND IDENTITY-SPECIFIC SEMANTICS

It has been argued in the preceding sections that the peculiarity of proper-name processing stems from a different organization in the semantic system. Recent experimental literature postulates that the semantics specific to an individual hold a related independent position in the semantic system. It has been shown that this may very well be motivated by the particular organization that token reference has with respect to type reference: in the first but not in the second the link between name and reference is unique. Proper names indeed hold together semantic features that are put together not for a category but, combined by chance, to form the description of a unique individual. With respect to these notions, neuropsychology proves again important in providing empirical support.

The case will be briefly reported here of C. B. (fully described in Semenza, Zettin, M., & Borgo, 1997), a patient suffering from brain trauma, with a lesion in the parieto-occipital area and of another small lesion in the anterior frontal lobe. At the time of referral the patient complained of anomia for proper names. Indeed he showed a severe problem in retrieving proper names and a very mild anomia for common names. He had, however, a more extensive range of problems than other proper-name anomics (a similar patient, whose disorder was, however, of developmental origin was also reported by Van der Linden, Brédart, and Schweich, 1995). First of all, unlike proper-name anomics he could not retrieve any other type of infor-

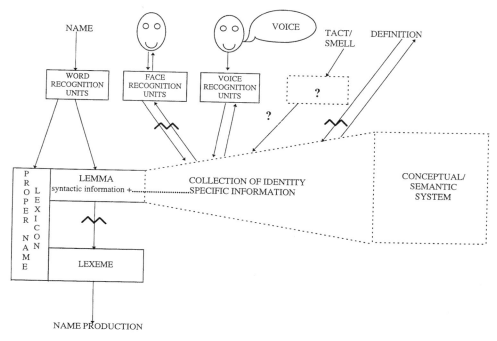

Figure 1. A model of proper-name processing and C. B's deficit.

mation about people he could not name. This happened with pictures and also with real people. He was not, however, prosopagnosic, because he could perfectly distinguish people he knew from people he did not know and scored within normal limits on a face-recognition memory test. Another peculiar feature was that he could not retrieve distinguishing information about a person when provided with other information about the same person. For instance, if provided with the cue "president of FIAT" he might say "very rich" (an easy guess) but not "white-haired." This contrasted with the fact that, when provided with proper names, C. B. could then immediately provide full information about the people he knew, including details unique to each individual. He could not, however, match spoken or written names of famous people—even those for whom he had demonstrated full knowledge—to corresponding pictures in a multiple-choice test. It thus seems that, for C. B., faces became disconnected from person-specific information. This specific type of information seems isolated from other types of information (see Figure 1). The only way of accessing it is via the corresponding proper names, which prove therefore, in full accordance with theoretical views, that they are simple pointers to the individual address in memory. This function of labeling individuals is the raison d'être of proper names. The increasing demands in social skills may have favored the evolution of a system dedicated to retrieve information about individuals, which at the same time, accounts for their well-known vulnerability.

REFERENCES

Bolla, K. I., Lindgren, K. N., Bonaccorsi, C., & Blecker, M. L. (1991). Memory complaints in older adults. Fact or fiction? *Archives of Neurology, 48,* 61–64.

Bruce, V., & Young, A. (1986). Understanding face recognition. *British Journal of Psychology, 77,* 305–327.

Burke, D. M., MacKay, D. G., Worthley, J. S., & Wade, E. (1991). On the tip of the tongue: What causes word finding failures in young and older adults. *Journal of Memory and Language, 30,* 542–579.

Burton, A. M., & Bruce, V. (1992). I recognize your face but I can't remember your name: A simple explanation? *British Journal of Psychology, 83,* 45–71.

Burton, A. M., & Bruce, V. (1993). Naming faces and naming names: Exploring an interactive activation model of person recognition. *Memory, 1,* (4), 457–480.

Butterworth, B. L. (1989). Lexical access in speech production. In W. Marslen-Wilson (Ed.), *Lexical representation and process.* Cambridge, MA: MIT Press.

Carney, R., & Temple, C. M. (1993). Prosopanomia? A possible category-specific anomia for faces. *Cognitive Neuropsychology, 10*(2), 185–195.

Cipolotti, L., McNeil, J., & Warrington, E. K. (1993). Spared written naming of proper nouns: A case report. *Memory, 1,* (4), 289–311.

Cohen, G. (1994). Age related problems in the use of proper names in communication. In M. L. Hurrmart, J. M. Wieman, & F. Nussbaum (Eds.), *Interpersonal communication and older adulthood: Interdisciplinary research.* Los Angeles: Jage.

Cohen, G. (1990). Why is it difficult to put names to faces? *British Journal of Psychology, 81,* 287–297.

Cohen, G., & Burke, D. M. (1993). Memory for proper names: A review. *Memory, 1,* (4), 249–263.

Cohen, G., & Faulkner, D. (1986). Memory for proper names: Age differences in retrieval. *British Journal of Development Psychology, 4,* 187–197.

Cohen, L., Bolgert, F., Timsit, S., & Cherman, J. F. (in press). Anomia for proper names following left thalamic infarct. *Journal of Neurology, Neurosurgery, Psychiatry.*

Crook, T. H., & West, R. L. (1990). Name recall performance across the adult life span. *British Journal of Psychology, 81,* 335–349.

Damasio, H., Grabowski, T. J., Tranel, D., Hichward, R., & Damasio, A. R. (1996). A neural basis for lexical retrieval. *Nature 380,* 499–505.

Damasio, A. R., & Tranel, D. (1990). Knowing that 'Colorado' goes with 'Denver' does not imply knowledge that 'Denver' *is* in 'Colorado.' *Behavioural Brain Research, 40,* 193–200.

Damasio, A. R., & Tranel, D. (1993). Nouns and verbs are retrieved with differently distributed neural systems. *Neurobiology, 90,* 4957–4960.

Fery, P., Vincent, E., & Brédart, S. (1995). Personal name anomia: A single case study. *Cortex, 31,* 191–198.

Flude, B. M., Ellis, A. W., & Kay, J. (1988). Face processing and name retrieval in an anomic aphasic: names are stored separately from semantic information about people. *Brain and Cognition, 11,* 60–72.

Frege, G. (1892). Uber Sinn und Bedeutung [On the sense and meaning]. In Patzig (Ed.), *Funktion, Begriff, Bedeutung* (pp. 40–65). Gottigen: Vandenhoek und Ruprect.

Goodglass, H., & Wingfield, A. (1993). Selective preservation of a lexical category in aphasia: Dissociations in comprehension of body parts and geographical place names following focal brain lesion. *Memory, I,* (4), 313–328.

Harris, D. M., & Kay, J. (1995). Know your face but I can't remember your name: Is it because names are unique? *British Journal of Psychology, 86,* 345–358.

Hittmair-Delazer, M., Denes, G., Semenza, C., & Mantovani, M. C. (1994). Anomia for proper names. *Neuropsychologia, 32,* (4), 465–476.

Jackendoff, R. (1983). *Semantics and cognition.* Cambridge, MA: MIT Press.

Katz, J. (1972). *Semantic theory.* New York: Harper & Row.

Kripke, S. (1980). *Naming and necessity.* Oxford: Basil Blackwell.

La Palme Reyes, M., MacNamara, J., Reyes, G. E., & Zolfagari, H. (1993). Proper names and how they are learned. *Memory, I,* (4), 433–455.

Levelt, W. J. M. (1989). *Speaking. From intention to articulation.* Cambridge, MA: MIT Press.

Lucchelli, F., & De Renzi, E. (1992). Proper name anomia. *Cortex, 28,* 221–230.

Luzzatti, C., & Whitaker, H. (1996). Johannes Schenck and Johannes Jakob Wepfer: Clinical and anatomical observations in the prehistory of neurology and neuropsychology. *Journal of Neurolinguistics, 9*(3), 157–164.

MacKay, D. G. (1987). *The organization of perception and action: A theory for language and other cognitive skills.* New York: Springer-Verlag.

MacNamara, J. (1982). *Names for things. A study of human learning.* Cambridge, MA: MIT Press.

McKenna, P., & Warrington, E. K. (1978). Category-specific naming preservation: A single case study. *Journal of Neurosurgery and Psychiatry, 41,* 571–574.

McKenna, P., & Warrington, E. K. (1980). Testing for nominal dysphasia. *Journal of Neurology, Neurosurgery, and Psychiatry, 43,* 781–788.

McNeil, J. E., Cipolotti, L., & Warrington, E. K. (1994). Accessibility of proper names. *Neuropsychologia, 32,* 4, 455–465.

McWeeny, K. H., Young, A., Hay, D. C., & Ellis, A. W. (1987). Putting names to faces. *British Journal of Psychology, 78,* 143–144.

Mill, J. S. (1843). *A system of logic* (10th ed. 1879). London: Longmas.

Miller, G. A., & Johnson-Laird, P. (1976). *Language and Perception.* Cambridge, MA: Harvard University Press.

Reason, J. T., & Lucas, D. (1984). Using cognitive diaries to investigate naturally occurring memory blocks. In J. E. Harris & P. E. Morris (Eds.), *Everyday memory, actions and absent-mindedness,* London: Academic Press.

Russel, B. (1905). On denoting. *Mind, 14,* 479–493.

Saffran, E. M., Bogyo, L. C., Schwartz, M. F., & Marin, O. S. M. (1980). Does deep dyslexia reflect right-hemisphere reading? In M. Coltheart, K. Patterson, & J. C. Marshall (Eds.), *Deep dyslexia.* London: Routledge & Kegan.

Saffran, E. M., Schwartz, M. F., & Marin, O. S. M. (1976). Semantic mechanisms in paralexia. *Brain and Language, 3,* 255–265.

Searle, J. R. (1969). *Speech acts.* Cambridge, UK: Cambridge University Press.

Semenza, C. (1995). How names are special: Neuropsychological evidence for dissociable impairment and sparing of proper names knowledge in production. In R. Campbell & M. Conwey (Eds.), *Broken memories, neuropsychological case studies.* London: Basil Blackwell

Semenza, C. (1996). Methodological issues. In G. Beaumont, P. M. Kenealy, & M. J. L. Rogers (Eds.), *The Blackwell dictionary of neuropsychology.* Oxford: Basil Blackwell.

Semenza, C., Nichelli, F., & Gamboz, N. (1996). The primacy effect in free recall of lists of common and proper names: A study on young, elderly, and Alzheimer disease subjects. *Brain and Language, 55*(1), 45–47.

Semenza, C., & Sgaramella, T. (1993). Production of proper names: A clinical case study of the effects of phonemic cueing. *Memory, I,* (4), 265–280.

Semenza, C., & Zettin, M. (1988). Generating proper names: a case of selective inability. *Cognitive Neuropsychology, 5*(6), 711–721.

Semenza, C., & Zettin, M. (1989). Evidence from aphasia for proper names as pure referring expressions. *Nature, 342,* (6250), 678–679.

Semenza, C., Zettin, M., & Borgo, F. (1997). *Person names and personal identification.* Unpublished manuscript.

Shallice, T., & Kartsounis, L. (1993). Selective impairment in retrieving people's names: A category specific disorder? *Cortex, 29,* 281–291.

Valentine, T., Brédart, S., Lawson, R., & Ward, G. (1991). What is a name? Access to information from people's names. *European Journal of Cognitive Psychology, 3,* 147–176.

Valentine, T., Brennen, T., & Brédart, S. (1996). *The cognitive psychology of proper names.* London: Routledge.

Van der Linden, M., Brédart, S., & Schweich, M. (1995). Developmental disturbance of access to biographical information and people's names: A single case study. *Journal of the International Neuropsychological Society, 1,* 589–595.

Van Lanker, D., & Klein, K. (1990). Preserved recognition of familiar personal names in global aphasia. *Brain and Language, 39,* 511–529

Wapner, W., & Gardner, H. (1979). A note on patterns of comprehension and recovery in global aphasia. *Journal of Speech and Hearing Research, 29,* 765–772.

Warrington, E. K., & Clegg, F. (1993). Selective preservation of place names in an aphasic patient: A short report. *Memory, I*(4), 281–288.

Warrington, E. K., & McCarthy, R. A. (1983). Category specific access dysphasia. *Brain, 106,* 859–878.

Warrington, E. K., & McCarthy, R. A. (1987). Categories of knowledge. Further fractionation and attempted integration. *Brain, 110,* 1273–1296.

Wittgenstein, L. (1922). *Tractatus Logico-philosophicus* (C. K. Ogden, Trans.). London.

Wepfer, J. J. (1727). Observationes medico-practicae de affectibus capituis internis & externis [Medical-practical observations of affections inside and outside the head]. Ziegler: Shaffausen.

Young, A., Hay, D. C., & Ellis, A. W. (1985). The faces that launched a thousand slips: Everyday difficulties and recognition people. *British Journal of Psychology, 76,* 495–523.

Part IV

Life Span Perspectives on Anomia: Clinical and Therapeutic Considerations

Considerable insight can be gained about the naming process and its components by looking at normal and aberrant patterns of naming at opposite ends of the age spectrum. Paula Menyuk, in the first of the three chapters of this section, deals with naming in children by first focusing on lexical acquisition in normally developing children. Here she traces the process of concept acquisition and the attachment of names to concepts. She next goes on to discuss naming problems in children who have known lesions or developmental brain abnormalities. In the third part of this chapter Dr. Menyuk examines naming problems in children who have specific language impairment (SLI), where firm evidence of a brain lesion is not easily established.

In the next chapter, Marjorie Nicholas, Christine Barth, Loraine Obler, Rhoda Au, and Martin Albert recognize that name retrieval is often a problem in normally aging people and describe longitudinal data on the naming of both objects and acts in elderly adults. Here they consider a number of theoretical models that have been proposed, such as the Transmission Deficit Hypothesis, which has particular bearing on difficulty in retrieving peoples' names. In their treatment of the breakdown of naming in Alzheimer's disease, they deal with both the phenomenology of the disorder and the theoretical underpinnings that relate it to the normal naming process.

In the final chapter of this section, written by Nancy Helm-Estabrooks, various approaches to the remediation of naming disorders are examined both from a historical viewpoint and in terms of their rationale. Dr. Helm-Estabrooks classifies these approaches into those that attempt a functional reorganization of the naming process, those that involve deblocking, and those that are driven by cognitive models. She both describes the methods and reviews the evidence for their effectiveness. We consider it appropriate to close this volume with a chapter that deals with the interface between theories of anomia and applications to individuals in need of remediation of this disorder.

Chapter 6

Naming Disorders in Children

Paula Menyuk

INTRODUCTION

The production of recognizable words is that aspect of language development that many people believe is the true beginning of the process. However, researchers have known for some time now that a great deal that is part of language development has already occurred before first words appear (Bloom, 1991; Menyuk, 1992). Infants have acquired, used, and understood some prosodic means (intonation and stress), facial expression, and gesture to communicate their needs and feelings to their caregivers for many months before words appear (Bates, O'Connell, & Shore, 1987). They have also begun the process of recognition of the speech sounds in their language at about 6 months (Kuhl, 1990) and recognition of patterns of speech sound sequences in the second half of the first year of life (Morgan & Saffran, 1995). They indicate comprehension of some aspects of the meaning of sequences of these speech sounds or words at 10 months (Menyuk, Liebergott, & Schultz, 1995).

The fact that infants appear to know a great deal about words and language before they say their first words supports current theory and research in language acquisition that point to the profound role that word recognition and production play in all other aspects of language development. This current theory suggests that word acquisition takes place in conjunction with syntactic and phonological development, and, therefore, plays a crucial role in these developments as well. Because of the role of lexical acquisition in language development, children with a naming disorder or, rather, children with lexical acquisition and retrieval problems will have difficulty in other aspects of language development. Studies of the language problems of these children certainly support this notion (Menyuk, 1993). These children also exhibit differences and delays in syntactic and morphophonological development as well as discourse difficulties. However, this chapter will focus on what is known about children's naming disorders, or lexical problems, per se.

ANOMIA: Neuroanatomical and Cognitive Correlates

The first part of the chapter will discuss factors that are said to play a role in the lexical acquisition of normally developing children. Theories of lexical acquisition and early patterns of development will be described. This provides a background for the discussions on children's lexical acquisition and naming disorders. Then studies of the lexical problems and naming behaviors of children with known lesion will be reviewed. Finally, the naming problems of children with suspected lesion or specifically language-impaired (SLI) children will be examined. These latter children are those who have been described as being normal in cognitive development but delayed or different in acquisition of some or all aspects of language development.

LEXICAL ACQUISITION IN NORMALLY DEVELOPING CHILDREN

Data that are currently available suggest that infants begin to recognize words at approximately 10 months of age; however, there are certain conditions that are required before they are able to do so. Word recognition, as reported in parent-kept diaries (Menyuk et al., 1995) and in experimental situations (Thomas, Campos, Shucard, Ramsey, & Shucard, 1981), begins with a very small set of words. In order to be recognized these words should be spoken by someone who is addressing the infant and not another adult. Recognition is significantly better when the word is spoken by a female, preferably the mother. These required conditions indicate that word recognition is very different in the young infant as compared to a somewhat older child and, certainly, an adult. They also indicate that the speech signal plays an important role in that recognition, hence the need for words being produced in a particular way by particular speakers. Despite the fact that there are restrictions on what is known about words at this age, certain fundamental strategies appear to begin to develop at this early age. First, the notion that a particular speech sequence is related to a particular object or event is developed. That is a crucial notion. Another crucial insight is that when these sequences are very similar they most likely are related to the same object or event. Lexical acquisition relies on the ability to recognize similar sequences as, in all likelihood, referring to similar objects and events, and observing that different sequences probably refer to different objects and events.

Early word production is also constrained by a number of conditions, but they differ somewhat from those observed with word recognition. Words begin to be produced later than they are recognized. The set of words recognized and produced overlaps to some extent but not completely. Thus, for example, infants express needs and feelings with a set of words referring to themselves ("me," "my," "mines") without necessarily recognizing these words when others produce them. Early word production is governed by both those factors that affect word recognition and other factors as well. The word must be recognized on some level before it can be produced but, in addition, the ability to map articulatory movements to match the perceptual representation of a word calls for additional abilities. Thus, early word production is slower but, also, it is thought to be much more approximate than word recognition. For example, the infant might not accept the sequence "tap" as referring to an animal with whiskers who says "meow" but might produce this sequence to refer to "cat."

Thus far we have talked about the need for the child to have an acoustic or phonological representation stored in memory for a small group of words, knowledge of what objects or events this small group refers to, and knowledge of how to approximate the acoustic or phonological representation stored with appropriate articulatory movements. Therefore, the infant's knowledge of words includes, at least, acoustic or phonological features, semantic or reference features, and articulatory features. There are other kinds of knowledge that the infant has early on, such as knowledge of what an object or an event is (Gibson & Spelke, 1983). They must have some strategy to retrieve a representation and compare it to a word heard to decide whether it is the same or different than those stored. They must be able to infer that if it is different it probably refers to a different object or event than those stored (Markman, 1991), and they have the ability to store and retrieve new representations. This is a formidable array of abilities, but this is just the beginning. To add to this array of abilities, the infant also has notions about the syntactic role of words in utterances (Pinker, 1987). Children as young as 2 years have been found to be sensitive to the syntactic environment in which novel words are introduced in terms of classifying words (Waxman, 1991).

What the infant, the young child, the older child, and adolescent know about the meaning of the *same* lexical items changes remarkably over the developmental years. Even simple lexical items such as *cup* change their meaning as the child matures. These changes in meaning have been described in varying ways. Initially, it is proposed, the child recognizes this lexical item as referring to an instance, perhaps their own cup. This has been described as *underextension*. Then the child observes that objects in the environment that have a certain appearance and are used to drink out of, are all referred to as *cup*. This generalization or so-called semantic feature development for words in the lexicon has been referred to as denotation because the word is now being used to stand for a class of objects and not, simply, to refer to an object.

It has been argued by some that classification of perceptual experiences, or concept development, is crucial to lexical development (Siegler, 1991). Others have argued that lexical acquisition is crucial for concept development (Vygotsky, 1962). There are discussions about whether the child develops the concept or category cup, and then recognizes and applies the word appropriately to all members of the category cup, or if the inverse occurs. That is, the child may acquire the word, then observe its application to many instances and only then come up with the concept of cup. Determination of the sequence in which concept or word acquisition occurs is important for understanding the problems of those who have difficulty in word acquisition and retrieval.

The meaning of the word *cup* develops further. One way in which it develops is that the word is recognized as being a member of a larger set (things that one eats and drinks with, china ware), and it is recognized as being the root of other compounds (coffee cup, tea cup). The first type of knowledge has been termed superordination and the second further subordination. Further elaboration of the meaning of the word occurs in terms of reference (the word can be both a noun and a verb) and multiple meanings of the same word develop (the cup you drink out of, and the cup you win in a competition). The word can also be used in metaphorical expressions ("my cup runneth over") in which the word has no direct reference to an object of a certain shape, size, and function. If the meaning of words is represented in memory as a set of features then, presumably, as the meaning for the word develops and changes, this set of features develops and changes. Table 1 describes the developmental changes that are said to occur in the meaning of words for objects. Somewhat different changes occur for words in different semantic fields and syntactic classes.

In addition to discussions about the direction in which concept development takes place, from word to concept or concept to word, there are different views of what types of features are stored for varying concepts and, therefore, one could argue, also, for either words and/or concepts in our lexicons. In a discussion of word meaning (Hirsh-Pasek, Reeves, & Golinkoff, 1993), the differing views of the type of features that are stored are reviewed. The classic view is that the features of a word or concept are necessary and sufficient for determining what the word refers to. For example, "triangle" is said to refer to an object that is closed, has three sides, and has angles that add up to 180 degrees, and is not applicable to any other shape without these features. The features compose a set or are conjunctive (Clark, 1983).

Others describe these conceptual or word features as probabalistic, and words are stored in terms of a set of probable features. Concepts or words that are "prototypical" fit the set of features most closely. For example, the

Table 1 Changes in the Use of Lexical Items for Objects over Development and Presumed Changes in Meaning

Use	Meaning
I Word used for a specific instance of an object	I To refer to a specific object
II Word used for a class of objects that are similar perceptually and functionally	II To denote a class of objects
III Word used to denote a member of a set of objects that fit into a superordinate class	III Base and Superordinate relations
IV Word used as a compound to further define object	IV Base and Subordinate relations
V Word used to denote more than one type of object	V Multiple meanings
VI Word used metaphorically	VI Meaning extended beyond denotation

word *bird* probably refers to an animal that flies and has wings. The word *robin* would be a prototype for the set of features that we have stored for the word *bird* (Rosch, 1975). The third feature theory, or exemplar view, suggests that what is stored are instances to which the concept or word applies (Medin & Schaffer, 1978). This allows for both overlap and distinction among objects to which the word applies. For example, bird can be applied to robins, owls, and ostriches because in our experience we have found that they are all instances of the category bird. The word is not applied to apples because apples are not instances of the category bird. In this sense the theory is disjunctive. These three theories of the nature of the features that are stored as the meaning of words have been verified to some extent by research with both children and adults. However, as with all experiments, the conditions under which the experiments are run have a direct effect on which theory seems to account most for the responses of subjects.

Given the current theorizing it is difficult to determine what the exact nature of the semantic knowledge about words that is stored in memory is, and, therefore, difficult to determine what the possible basis of the difficulty might be in children and also adults with naming disorders. This theorizing suggests it might lie in a bundle of defective conjunctive, probabilistic or disjunctive features, or unclear concepts, or some combination of these, about the meaning of a word. As will be discussed, the conditions under which naming behavior is elicited can have an effect on the results and, therefore, on our understanding of the relation between brain lesion, naming behavior, and the underlying causes of the behavior.

As noted earlier, words begin to be recognized by about 10 months of age. By 18 months children, are, on average, comprehending more than 100 words and producing 50 words. Lexical knowledge continues to grow at an amazing rate over the early childhood period (from 3 to 8 years), and, of course throughout life. Much of this early rapid growth occurs across and within different semantic fields. For example, children begin to acquire prepositions of location after they acquire nouns for common objects. They acquire nouns for common objects before they acquire nouns for less common objects. Within the field of preposition they acquire some prepositions before others (*in* and *on* before *under* and *behind*) (Menyuk, 1988). Cognitive, linguistic, and experiential factors are said to play a role in this observed sequence of acquisition of lexical items. Thus, in the acquisition of prepositions of location the infant must be able to observe the difference between objects being on something versus under something, and they have this ability at a very early age (Quinn, 1994). They must then also understand that although on is absolute (an object is on another or it is not) under is a relative term (an object might be under one object and on another). In addition, the child must learn what terms are used in the language to describe these relations among objects, and languages differ to some extent in terms of the nature of this marking of relations. For example, some languages mark the *on* of covering differently than they mark the *on* of on top of. In terms of common versus rare words the effect of frequency of usage obviously plays a role in the sequence of acquisition of words.

This great proliferation in word knowledge has been described as the ability to use the strategy learned earlier (if it is different it probably has a different label) in a remarkably rapid way. A recent study (Anglin, 1993) of lexical growth from first grade, when most children are about 6 years, until 5th grade, when most children are about 10 years, estimates that children have about 10,000 words at first grade, 20,000 at third grade, and 40,000 at fifth grade. A large part of the extremely rapid growth from third to fifth grade is due to children's knowledge that new words can be derived from old words by compounding, adding prefixes, suffixes, and infixes to root words. Some of this latter remarkable growth is due to education and the acquisition of reading. Written words provide stable evidence about the derivation of words: they can be examined at length and analyzed. These developments over the school years point to the child's ability to see relations among words on semantic, phonological, and syntactic grounds, which adds enormously to their lexical growth. The ability to see relations would, of course, require storage of semantic features in a way that would allow these relations to be observed.

Thus far, theory and research indicate that knowledge about words includes their phonological, morphophonological, syntactic, and semantic properties and the contexts in which they are used, and all these properties and uses of words begin to be acquired at a very early age. Less clear is what is included in these semantic properties. They may be bundles of defining or probabilistic features or theories about the essence of the meaning of the words or all of the above. These theories about the meanings of words may be based on experiences within a particular sociocultural group. It is possible that the semantic properties of lexical items include both features and theories and may, in addition, include personal theories about words. For example, *cat* may include among its semantic properties "fearsome animal" if an individual has a particular loathing for cats.

In addition to there being many theories about lexical storage, there are also many theories about lexical retrieval. These varying theories, however, all point to the use of context, both linguistic and extralinguistic, to access an area of the lexicon, and then use of semantic and phonological features to select the appropriate lexical item. Many studies of word recognition use pictures of single objects or events and require that the child recognize the word by pointing at the appropriate picture. This is also the case with many tests of vocabulary recognition. In a similar fashion, confrontation naming requires that the child name a picture of a single object or event. These tasks are different from the task involved in recognizing and producing words in connected discourse. As stated, most current theories of the latter processes take into account the role of context in both recognizing and producing words. In terms of recognition, context is said to allow prediction so that minimal cues are needed to recognize a particular lexical item (Marslen-Wilson & Tyler, 1980). For example, in the sentence "The boy is sleeping in the b___," the word *bed* would be predicted and then confirmed by the initial sound in the word, and the semantic features of the word *bed*. In terms of production, the kinds of speech errors made by both children and adults indicate that planning for production occurs over the phrase and the clause (Fromkin, 1993). For example, speech errors such as "tips of the slung" (slips of the tongue) across the phrase and "Did you stay up late very last night" (Did you stay up very late last night) across the clause occur. There are occasions when a word alone must be recognized. This is when the context is ambiguous. Under these circumstances some look-up procedure must be used to both identify the phonological sequence that is the word and the semantic properties of the word. This occurs rarely in discourse, and when it does requests for clarification can be made.

All of these factors that affect lexical comprehension and production should be kept in mind as the research on naming problems of children with known and suspected lesions is reviewed in the next two sections of

the chapter. The fact that what is known about lesioned children's difficulties is largely based on the results of confrontation-naming experiments
needs to be kept in mind.

LEXICAL PROBLEMS OF CHILDREN WITH DIAGNOSED LESION

Lesion can occur prenatally, perinatally, as well as postnatally. There are
several questions about the effects of lesion on language development in
general, and lexical acquisition, in particular, that researchers have asked.
For many years the first and only question asked was, What effect does age
of occurrence of lesion have on language development in children? For
some time it was held that lesions that occurred early in life had no long-
term effects on language development due to the plasticity of the nervous
system (Lenneberg, 1967). However, some longitudinal studies of early lesion came up with mixed results concerning later effects. Some studies
found long-term effects while others did not. Also, more modern research
began to indicate that early damage might lead to permanent problems in,
at least, some aspects of language. In addition, it was never clear how
young the child had to be to show no permanent damage from early lesion,
and there continue to be lengthy discussions about this issue.

Some researchers questioned the validity of studies that made claims
about early damage having no long-lasting effects because, these researchers suggested, the subjects, used in the studies that examined the
question of long-term effects of lesion probably had suffered varying types
of damage (Wittelson, 1977; Woods & Carey, 1979). Therefore, the persis-
tence of the effects might be due to only certain types of lesion and not others. In an early review of such research, Robinson (1981) stated that when
the types of lesion suffered by children were similar to those found in
adults, then the difficulties encountered by the children were similar to
those of adults.

These differences in findings that were the results of research in the 1970s
and 1980s led to a further examination of the all-or-none position that had
been held previously, and to the further question, Does site and size of lesion have a differential effect on language development? Present-day research, in which brain-imaging techniques are used to determine site and
extent of lesion, allow researchers to ask these questions with some accuracy and to obtain more complete answers. Current research is also concerned with the correctness of the statement made by Robinson, that similar sites and size of lesion in children lead to language behaviors similar
to that of adult aphasics. Thus, age remains a persistent question, but it is
now framed in a different and possibly more insightful manner. That is, the

question has become, In terms of processing language, what is fixed and what is not fixed in the nervous system of the young infant, and when is it fixed?

Both the confrontational naming and lexical acquisition of children with focal brain injury have been examined in recent studies. As stated, in present-day research with children with focal brain injury the site, left or right hemisphere, anterior or posterior, and often size of lesion are known. This, as stated, allows the above posed questions to be answered somewhat more accurately, although, as will be seen, not totally conclusively. Since it is very difficult to know exactly what is happening in the nervous system when recovery takes place, an argument can and has been made that early fixation of function in the nervous system can be altered by cognitive and behavioral development, not change in neurological specialization, when recovery from lesion occurs (Wittelson, 1987). Arguments such as lesion-related release of pre-existing functional potentials (in other words change in neurological specification) have also been made (Bullock, Liederman, & Todorovic, 1987). Because of the behavioral developmental changes that occur in children, it is difficult to know if either or any argument is correct.

Data obtained on the linguistic and nonlinguistic abilities of hemispherectomized children by Dennis indicated that those who have their left hemisphere removed have great difficulty with certain syntactic processing. (Dennis, 1980a). Research on these children's word finding indicated that, although the children studied had no word-finding problems, the child whose right hemisphere was removed was better than the children with their left hemispheres removed on conceptually based semantic tasks (Dennis, 1980b). The distinction is made by this researcher between automatized word retrieval and comprehension and production of words in connected discourse tasks, and the suggestion is made that there is a dissociation between these abilities. This research points up the importance of the task conditions in determining naming problems in children. That is, some tasks call for automatic processing, whereas others require lexical accessing through phonological and/or semantic and/or syntactic processing.

Several studies have been carried out examining confrontation naming in children who have suffered left or right brain lesions. One of the earliest of these (Woods & Carey, 1979) found that those who suffered left brain lesion after age 1, but not before, were significantly less accurate on a particular naming test than normally developing control subjects. Another study (Kiessling, Denckla, & Carlton, 1983) found that children with either left or right brain lesion had lower scores on another test of naming than did their sibling controls. There were no differences in performance by the left- versus the right-lesioned children. Still another study (Vargha-Kadem, O'Gorman, & Watters, 1985) examined naming on the test employed in the first mentioned study in a group of children who had suffered lesions prenatally, early postnatally, and later postnatally. It was found that all left-lesioned children regardless of time of lesion were impaired in accuracy of naming when compared to control subjects. Children who had had a right lesion early postnatally were also impaired in accuracy of naming, but the other groups of right-lesioned children were not. There was a significant negative correlation between age of lesion and naming accuracy. Aram, Ekelman, Rose, and Whitaker (1985) used still another test to examine naming in a group of 8 right-lesioned and 8 left-lesioned young children. The children with left lesion suffered damage at ages ranging from 1 month to 6 years and those with right lesion at ages ranging from 2 months to 3 years. Each child was matched with a non lesioned peer on sex, age, socioeconomic status (SES), and race. When testing took place the left-lesioned children ranged in age from 2 to 8 years and the right from 2 to 6 years. It was found that all but one of the left-lesioned children scored below their controls, but the differences were not significant. Of the eight right-lesioned children, four scored lower than their controls, three scored higher, and no data were obtained from one child.

Unfortunately, the conflicting results of these studies can be the result of multiple confounding factors. It is possible that the use of different tests bring about varying results, hence the differences among the studies in terms of the effect of left versus right lesion. Age at time of lesion may also have an effect on whether or not right-lesioned and left-lesioned children continue to have problems in naming accuracy, and on whether or not there is a significant effect on naming accuracy regardless of side of lesion. Age at testing and naming accuracy are probably related, and naming accuracy, rather than some other measure of naming ability, might be affected to a lesser or greater extent by the particular lexical items used in each of the studies; that is, the particular phonological and semantic properties of the lexical items used in each test.

To overcome some of these confounding factors a study was carried out with a sizeable number of right- (13) and left- (19) lesioned children (Aram, Ekelman, & Whitaker, 1987). Except for the five children who sustained prenatal lesion, documentation of normal development prior to lesion was available, thus eliminating the presence of other complicating factors. The left-lesioned children had incurred damage at times ranging from prenatally up to $11\frac{1}{2}$ years and were tested at ages ranging from 6 to over 14 years. The right-lesioned children suffered damage at ages ranging from prenatally up to almost 16 years and were tested at ages ranging from approximately $5\frac{1}{2}$ years to $16\frac{1}{2}$ years. These subjects were matched to controls on age, sex, SES, and race. Subjects were not individually matched on measured IQ but the left-lesioned and control groups did not significantly differ from each other. The full scale IQs for the right-lesioned group, however, were significantly lower than those of their control group. This is an important factor to control for because vocabulary knowledge, as on the Peabody Picture Vocabulary Test, and measured IQ, as on the Wechsler Intelligence Test, are highly and significantly correlated with each other.

The children were given two tests of naming. The first test, The Word Finding Test (Wiegel-Crump & Dennis, 1984) included items in several semantic fields (animals, food, clothing, household items, and actions), and words were requested with three cuing conditions (rhyming, semantic descriptors, and visual presentation). Thus, the first test was not simply a confrontation-naming test. Both latency of response and errors were recorded. Presumably, all the words employed are available to 4- to 6-year-old children. The second test was the Rapid Automatized Naming (RAN) test (Denckla & Rudel, 1976). The youngest subjects in the study were not able to name letters, numbers, or both but were able to name colors and objects. Latency of response and errors were again recorded. Comparisons were made between lesioned subjects and their controls rather than between left- and right-lesioned subjects because the latter varied from each other across important dimensions, such as age and measured IQ. Results for each test are reported separately.

On the word-finding test, the left-lesioned children were significantly slower than their controls in their overall performance, and when semantic and visual cues were provided. Rhyming cues were difficult both for left-lesioned children and their controls. Left-lesioned experimental subjects were significantly less accurate than their matched controls both in overall performance and when rhyming cues were provided. Right-lesioned subjects were faster in response in general than their controls, and significantly so when rhyming cues were provided. Right-lesioned subjects were significantly less accurate than their matched controls overall, but there were no significant differences between these two groups with particular cuing conditions. It was also found that the semantic category of the lexical items had a significant effect on the latency and accuracy of response only for the left-lesioned group. Finally, associative responses that were out of the semantic field of the target item but related to the item in a functional way were twice as frequent for left-lesioned children as for any other group. Over half the errors with the rhyming condition were no responses, except for the right-lesioned subjects who tried to match the sound sequence, and the fewest number of errors were made with the visual or picture cues except for the right-lesioned subjects, who tended to make more errors than the other groups. With the RAN the left-lesioned subjects were significantly slower than their controls in all categories, although there were no significant differences in latency of response between right-lesioned subjects and their controls. There were differences in latency of response to various categories. Numbers and letters were named most rapidly, then colors, and finally objects.

The results of this study begin to answer some of the questions about the effect of site of lesion on the lexical accessing of children. The researchers point out that the overall pattern of responses by the left-lesioned and right-lesioned subjects was similar to that of the normal subjects on whom the test had been normed. However, left-lesioned children were slower and less accurate than their controls, they tended to be even slower in the rhyming condition, although not significantly so, and they made many more associative errors when given semantic cues. If, as has been proposed by a number of researchers, lexical assessing for word production involves conceptual categorization and phonological realization, the left-lesioned subjects in this study were both slower and less accurate in utilizing both these accessing routes. But, as pointed out by the researchers, they did not exhibit the kinds of naming problems shown by adult aphasics. Perhaps they did not do so because the conditions for producing such errors were not present in the particular test situation.

The right-lesioned children were faster in responding than their controls, they made more errors in the visual condition of the word-finding test, and produced more visual association errors than did the other groups. Finally, in the rhyming condition, they attempted to match the disyllabic sequence presented rather than simply not responding, as all other groups of children did. Unlike their "impulsive" performance on the first test, they were slower than the control subjects in responding on the RAN where visual cues were always present. These two findings together point to visual-processing difficulties, as the researchers conclude. These results, however, may also point to possible reliance on syllabic prosodic information to categorize words, presumably an ability that develops in the right hemisphere. At any rate, this study indicates that early left lesion and right lesion affect lexical accessing in different ways, and that the effects of early lesion persist into these older ages.

Parenthetically, because the subject numbers were small, the researchers add that it appeared to make no difference whether a lesion occurred before or after the age of 1 year. There appeared to be an effect of site of lesion on left-lesioned children's performance; children with subcortical involvement showed the greatest latency and number of errors in response. Finally, because the left-lesioned children had no greater problems with repetition of nonsense syllables of increasing length than controls, their lexical retrieval problems could not be attributed to articulation difficulties. However, a distinction must be made between articulation difficulties and phonological realization rules in lexical retrieval. The former are difficulties that may involve only peripheral commands to the articulators, whereas the latter, presumably, involve more central processing as well as peripheral commands.

Almost all the studies that have examined lexical difficulties in lesioned children have focused their attention on naming problems or production. Almost all the studies that have been discussed here have examined word production in children with focal brain injury. Equally important questions concern (1) the effect of lesion on word comprehension, and (2) the relation between comprehension and production of words in lesioned children. In a study cited previously (Aram et al., 1985) lexical comprehension as well as production was measured. The children were given the Peabody Picture Vocabulary Test (PPVT). It was found that all groups of children (left-lesioned and controls, right-lesioned and controls) except the controls for the left-lesioned group scored higher on comprehension than production tests. The left-lesioned children scored significantly more poorly than their controls on both the comprehension and production measures. The right-lesioned children did significantly more poorly than their controls only on the lexical comprehension test. This difference may be either a cause or an effect because of the lower measured IQ of the right-lesioned as compared to the left lesioned children, and the significant relation known to exist between measured IQ and lexical comprehension.

A further study of the relation between lexical production and comprehension was carried out by Eisele and Aram (1993). In reviewing the literature on this topic the researchers pointed to some very interesting findings. They reviewed some developmental research on lexical acquisition in unilaterally brain-damaged children, which will be discussed next. The results of this research indicated to these investigators that lexical comprehension is less lateralized than lexical production and that there is a differential role of left- and right-hemisphere damage on the development of expressive and receptive lexicons. They further suggest that the lexical comprehension difficulties of right-lesioned subjects can be separated from the syntactic difficulties of left-lesioned subjects, as evidenced by their performance on the Token Test. The left-lesioned subjects do significantly more poorly in overall performance on this test than do right-lesioned subjects. However, this test is one in which there must be heavy reliance on knowledge of lexicon as well as syntax in order to perform well, and a lexicon that is heavily weighted toward adjectives (colors and shapes) as well as prepositions (of space and time). These are lexical items that are not the earliest acquisitions. Furthermore, current theory and research on lexical acquisition point to the simultaneous storage of semantic properties and syntactic context properties of lexical items. These results with differing tests, some of which only require lexical retrieval and others which require both lexical and syntactic knowledge, again raise the question of how the particular conditions of the experiment can affect results.

The subjects in the study (Eisele & Aram, 1993) were 21 children with unilateral left lesion and 12 with unilateral right lesion and 16 control subjects. The mean age at the time of lesion for the left-lesioned subjects was 2 years 3 months and for the right-lesioned subjects 1 year 8 months. The range of ages at which lesion was suffered was from prenatally to 11 years 7 months for the left-lesioned children, and prenatally to 9 years 8 months for the right-lesioned children. All children were given the PPVT and the Expressive One Word Vocabulary Test. When available, multiple test scores on both tests were used. In addition the children's measured intelligence was periodically tested using the varying forms of the Wechsler Intelligence Scales that are appropriate for the child's age.

The results were as follows. The left-lesioned subjects' performance on the naming test was comparable to that of the control children but poorer than that of the controls on the PPVT. The right-lesioned subjects scored lower than both the controls and the left-lesioned subjects on both the PPVT and the naming test. The scores of the three groups on the two lexical tests are shown in Figure 1. What was surprising about the results is that lexical comprehension was poorer than lexical production in both lesioned groups. The control children performed in a more expected man-

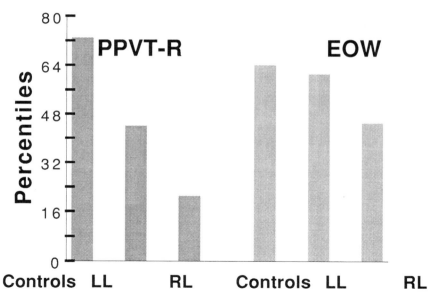

Figure 1   Word comprehension on the Peabody Picture Vocabulary Test (PPVT) and word production on the (EOW) test scores of lesioned children and controls. (From *Differential effects of early hemisphere damage on lexical comprehension and production,* 1993, pp. 513–523, J. Eisele & D. Aram, *Aphasiology, 7.* Reproduced with permission. All rights reserved.)

ner; their comprehension percentile score was better than their production score. In a discussion of the two tests, the researchers point to the possibility that the naming test was relatively easier than the comprehension test. This certainly might have affected the results. The nonlesioned groups might have found the vocabulary of the PPVT well within their competence and, thus, were relatively more competent in comprehending than in naming. The effect of lesion on lexical acquisition might much more severely delay the acquisition of the vocabulary used on the PPVT than that used on the Naming Test. This might have brought about the result that the two lesioned groups did better in naming than in comprehending, and blur the usual order of difficulty between naming and comprehension by these two groups. Using exactly the same test, or at least the same words, for the two tasks might overcome some of these difficulties.

All three groups showed a positive significant correlation between the naming test and the PPVT. The right-lesioned group scored significantly lower on IQ performance measures than did the left-lesioned and control groups. There was a significant correlation between IQ and lexical performance only for the left-lesioned group. The noncorrelation between right-lesioned subjects' lexical performance and IQ suggests that their relatively low comprehension performance cannot be attributed to a lower measured IQ, a possibility that was touched upon earlier. The researchers conclude that there is a significant right-hemisphere involvement in the acquisition of lexical knowledge, and, in particular, comprehension of word meaning. Given these results, it appears that when right-lesioned and left-lesioned children are compared on lexical performance, the right-lesioned children also do worse in naming as well as lexical comprehension. Furthermore, when only the subsamples of the left- and right-lesioned children that could be matched on age were compared, the results were quite similar. The right-lesioned group was more impaired than the left-lesioned group on both production as well as comprehension measures.

Tests of lexical comprehension and production may introduce some of the confounding variables we have talked about above. Longitudinal studies of vocabulary acquisition in children with focal brain damage eliminate some of the problems that particular test situations or experimental conditions introduce. However, they have implicitly some of the difficulties in interpretation that were discussed previously. It is not always clear whether developmental patterns of lexical acquisition present a clear picture of the effect of site of lesion on acquisition of lexical knowledge. Recovery from initial difficulties might be accounted for by a number of reasons that have equally to do with equipotentiality of sites in the nervous system, or the greater possibility for developmental changes to overcome some nervous system limitations.

The research on the effect of site of lesion on language development suggests the initial equipotentiality of the two hemispheres for the development of language. For example, Feldman et al. (1992) found that there were delays in early development of language with either right- or left-hemisphere lesion suffered prenatally. Some of the earlier research on lexical processing points to the effect of either right- or left-hemisphere lesion on lexical development, but the effect is presumably upon different aspects of processing: comprehension versus production. A question that arises from both types of research (developmental and experimental) is, Are the two hemispheres equal in terms of processing and storage of lexical information at birth or does each hemisphere deal with only certain aspects of lexical processing and storage from birth onward? This latter possibility seems to best explain the results of one recent developmental study.

Thal et al. (1991) studied the early lexical development of 27 children with focal brain damage. The children were aged 12 to 35 months and had suffered lesions prenatally or within the first 6 months of life. The population of 27 children was composed of 10 children studied longitudinally, on whom more than one data point on lexical development was available, and 17 children on whom one data point was available over the age range studied. There were several aspects of the effects of focal brain damage on lexical development that were examined in the study: (1) the effect of lesion size, (2) the effect of site of lesion, (3) the effect on lexical comprehension versus production, and (4) (for the first time) the effect on closed class words (grammatical function words) versus content words. Lexical development was assessed by use of the MacArthur Scales of Lexical Development, a parental check list that has been found to very reliably assess lexical development (Dale, Bates, Reznick, & Morisset, 1989).

The findings of the study contribute a great deal to our understanding of the effect of focal brain damage on lexical development, and, of course, leave us with many more questions than answers. Overall, the lesioned children gave clear evidence of delays in production and comprehension of words. To look at the effect of size of lesion children were divided into two groups, those with lesion in only one lobe and those with lesion across more than one. It was found that size had no significant effect on the presence, amount, and type of delay that was observed. Children with middle-sized lesions were more delayed than children with more extensive lesions. The researchers speculate that children with more extensive lesions may switch to the undamaged hemisphere earlier for language processing, and, in this way, compensate better for damage. Concerning side and location of lesion, it was found that left-lesioned children tended to be slower in development of productive vocabulary than right-lesioned children, but there were no significant differences between the groups. There was, however, evidence of slower development in the later phase of development (17 to 35 months), suggesting that left-lesioned children are recovering at a slower rate than right-lesioned children. Also, a so-called vocabulary "spurt" takes place in many normally developing children at around 18 months. This may have emphasized differences in lexical acquisition rates between right- and left-lesioned children. Comprehension of words seems to be more affected by right than left lesions, although, again, there were no significant differences. The development of verbs (referred to as "predication") tended to be slower in left- than right-lesioned children, and the use of closed class words more prevalent in the right- than the left-lesioned group.

The effect of site of lesion was examined further by comparing those children with lesion in the left posterior cortex and those without damage to the left posterior cortex. Expressive impairments were more severe in those with left posterior damage then in those without, but overall there was no significant difference between groups. However, differences between groups became more evident and significant when children at the later age period were compared; that is, when there was an increase in vocabulary acquisition in the period from 17 to 35 months. Left posterior damage was not a good predictor of comprehension difficulties nor of comprehension–production dissociations nor of verb acquisition. Finally, the proportion of use of closed class words was lower in children with left posterior damage than in children without such damage. The percentile differences between children with and without left posterior damage in expressive vocabulary at the later age period, and proportion of use of closed class words are shown in Figure 2.

The careful and tentative conclusions reached by the researchers have bearing on the questions raised initially about place, size, and age at time of lesion on lexical development. There was no linear relation in terms of size of lesion and magnitude of deficit. In fact, according to the researchers, the greater the size of lesion, the more rapid might be the recovery due to earlier use of compensatory structures. Children with left-hemisphere lesion were more delayed in production of words than those with right-hemisphere lesion, and this really became evident as vocabulary began to expand (at ages 17 to 35 months). Children with focal brain damage were unlike adult aphasics in that anterior lesions did not lead to production deficits, and left posterior lesions did not lead to comprehension deficits. There was weak support for right-lesioned children showing greater deficits in comprehension and left-lesioned children in production, and the latter having greater delays in the use of closed class vocabulary.

A comment here is necessary on possible differences in the target language affecting use of closed-class vocabulary by children with lesion. In a study of language development of children with left-hemisphere lesion acquiring Hebrew, it was found that the formal subsystems of the language

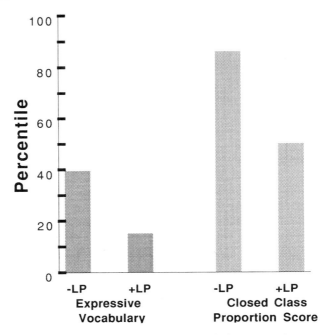

Figure 2 Expressive vocabulary and closed class proportion score
of children with and without damage to left posterior (LP) cortex.
(Adapted from Thal et. al., 1991.)

(morphosyntactic categories and relations) were much more preserved
than those that are meaningful (lexical or semantic-relational) (Yonata Levy,
Hebrew University, personal communication, February, 1992). This may
have to do with the differences between the richness of the morphology in
Hebrew versus English and the role that this morphology plays. Evidence
in support of this hypothesis, that the structure of the morphology might
have an effect on amount of preservation of closed class words, was found
in a study of the development of morphology by SLI Hebrew-speaking children as compared to SLI English-speaking children (Dromi, Leonard &
Shleitman, 1993). Although the use of closed class words by left-lesioned
children was delayed in the study of the English-speaking children reported, this was not the case with Hebrew-speaking children. Children with focal lesion learning Hebrew showed a higher *proportional* use of closed class
vocabulary on the whole, than do normally developing children.

As stated initially, the Thal et al. (1991) study provides a great deal of information about the lexical development of a comparatively large number
of children who have suffered focal brain damage. However, the answers

to the questions posed initially are still not available. Site and size of lesion appear to have an effect, but it is not always clear what the effect is and whether it continues over the subsequent years. This latter question has not been addressed as yet. Both the short and long term consequences of early lesion on lexical processing or naming problems are not clearly understood. In a review of studies on the effects of focal brain damage Bishop (1993) refers to many of the problems in the earlier research that were discussed, such as size of population, nondefinition of type of injury or site of injury, variable and, in some instances, simply observational methods of determination of problem, statistically suspect interpretation of data, and so on. The studies reviewed here seem to have overcome some of these methodological problems, and have begun to introduce the important notions that lexical processing is composed of subprocesses (comprehension and production), and that these subprocesses are affected by the properties of the particular lexical items being produced or comprehended (closed vs. content words). These issues are worthy of being pursued in further research, but problems with interpretation will still remain. The bases of these problems may lie in the lack of knowledge about the interaction of cognitive and social development with maturation of underlying biological systems.

NAMING PROBLEMS IN SPECIFICALLY LANGUAGE-IMPAIRED CHILDREN

The children to be discussed now are those who have been given various labels, including minimally brain damaged. They are children who rarely have had brain imaging techniques used with them, although they have frequently had neuropsychological examinations. Therefore, in most studies there are no data on the site and size of hypothesized brain damage but a great deal of data on their measured intelligence, cognitive, and linguistic abilities. Their presenting symptoms are, initially, a delay in linguistic development but no marked delay in cognitive development, although their measured IQ may frequently be at the low end of the normal range (Menyuk, 1993). Because measured IQ is largely based on language abilities, including so-called performance tasks, it is not surprising that there is a tendency for a number of these children to cluster at the low end of the normal range. Later in early development, they exhibit particular difficulties in acquiring certain aspects of language, in particular morphosyntactic aspects (Leonard, 1991) and syntactic aspects (Menyuk, 1993). However, word-finding difficulties have also been found in this population, and, as stated initially, if recent theorizing about lexical acquisition is correct, these children's syntactic difficulties may be related to limitations on their lexical knowledge or the interaction of the two aspects of language (Pinker, 1984). These limitations may lie in the syntactic properties associated with lexical items or in the phonological features that represent these items or both. In this section of the chapter the data on the lexical development of these children will be discussed first and then their behavior in word-finding studies.

In a longitudinal study (Curtis, Katz, & Tallal, 1992) the language development of SLI children and normally developing language age-matched peers was examined. The 28 SLI children had a mean age of 4 years 4 months at the beginning of the study, and their language age peers had a mean age of 2 years nine months. The children were studied over a 5-year period, and it was found that the two groups were similar both in the point at which they achieved mastery of structures and in their overall patterns of acquisition. The researchers suggest that the problems of the SLI children were not representational in nature but, rather, related to processing problems. One might conclude from these findings that knowledge about the properties of lexical items that is stored by these children is not remarkably different from the knowledge achieved by normally developing children. These processing problems bring about delays in acquisition but do not change the structures themselves. There is no easy way, if any, to determine whether or not this is the case. Again, it may be a matter of interpretation of the data obtained or there may be differences among SLI children, some of whom do represent structures in different ways and some who do not. It is the case that research indicates that at least some SLI children continue to have language difficulties as they reach the adult years (Tomblin, Freese, & Records, 1992). Therefore, these children do not completely recover from the language problems that they have, much like frankly lesioned children.

Lexical development is delayed in SLI children but the delay may be limited to word production. There are children who exhibit a marked delay in word production but have no difficulty, apparently, in word comprehension. These children have been called late talkers or children with expressive language delay. There is a great deal of variation among children in the rate at which they acquire a lexicon, both productive and receptive. However, the marked delay in these children of production and not reception continues on into the third and possibly fourth year of life, which is unusual (Rescorla & Schwartz, 1990). This delay in object naming is also found in use of symbolic gestures, and this combination of delay is said to be similar to the findings with some brain-damaged adults (Bates & Thal, 1991). Although children with expressive language delay are presumably at an age level in language comprehension they do exhibit other kinds of language delay, primarily in the morpho-syntactic aspect of language (Paul & Alforde, 1993), and in aspects of symbolic play (Rescorla & Goossens, 1992). The former findings seem quite similar to those reported in the developmental study of infants with defined left-hemisphere lesion (Thal et al., 1991).

Two other interesting aspects of early lexical acquisition in SLI children have been examined. Verb particle ("Put *on* your coat") and preposition acquisition ("Put the book *on* the table") were examined (Watkins & Rice, 1991) in a group of 4- and 5-year-old SLI children, and matched language age and chronological age peers. It was found that the SLI children, in addition to their morpho-syntactic difficulties, had difficulty in acquisition of these other grammatical form classes. In addition, other studies have found that verb learning may be particularly difficult for SLI children (Kelly & Rice, 1994; Oetting, Rice, & Swank, 1995). The researchers suggest that acquisition of these classes of words requires both semantic knowledge and also grammatical analysis of the distribution of these forms in utterances, perhaps a left-hemisphere processing ability. In another study (Rice, Buhr, & Nemeth, 1992), the fast mapping skills of these children was examined. Fast mapping (Carey, 1978) is that ability to very rapidly learn new lexical items that was discussed in the first section of the chapter. This ability accounts for the vocabulary spurt that occurs after about 18 months and continues to account for the very rapid growth of vocabulary over the school years. In this study an experiment was designed to expose the children to lexical items that mark objects, actions, attributes, and affective states, that is, content words. It was found that the SLI children scored lower on the number of words they could acquire than a language age-matched group and a chronological age-matched group. They had particular difficulty with object and attribute names. The investigators suggest that the SLI children have a restricted ability to comprehend new words, perhaps an early right-hemisphere processing disability. The findings point to possible both left- and right-hemisphere processing difficulties.

Overall findings that suggest similarity between SLI children and right-lesioned children are hard to come by. However, there may be SLI children who have, primarily, receptive language difficulties, including difficulty in comprehension of words and less expressive difficulty. Instances of this pattern of deficits occurring are much rarer than the inverse but such instances do occur to some extent. In a recent study of the reading abilities of SLI children (Menyuk et al., 1991), it was found that among the 23 children who were labeled SLI there were a few whose receptive language age was below that of their expressive language age. The criteria for inclusion in the SLI group were scoring 6 months or more behind their chronological age on receptive language tests and at least 12 months behind on lan-

guage-production tests. Thus, these few children were at least 1 year behind on their receptive language abilities. At 60 months, these children
were on average 12 months behind on both expression and reception of
language tests. Among a group of children who were classified as possibly
at risk for reading problems were those who gave early evidence of a language problem but who, at 60 months, were functioning in the normal age
range on most of the tests given on entrance. They certainly did not meet
the SLI criteria. With these children there was a greater prevalence of a receptive language disability. At 60 months they were 6 months above age
level on expressive language tests but at age level on receptive tests. These
results are shown in Figure 3. Furthermore, the SLI children's mean age
score on the PPVT was 51.1 months and on the Gardner (One Word Picture
Vocabulary Test) 53.7 months. The same scores for the at-risk group were
61.2 months for the PPVT and 71 months for the Gardner. This comparative difference between the SLI and at-risk children's production and reception scores looks somewhat similar to a finding in a previously cited
study (Eisele & Aram, 1993) that used the same lexical measures. In that
study right-lesioned children had comparatively greater deficits in comprehension than in production, although, unlike the subjects in this study,
they did more poorly on both tests than the left-lesioned children as well
as the control subjects. Both sets of results on word finding may be a func-

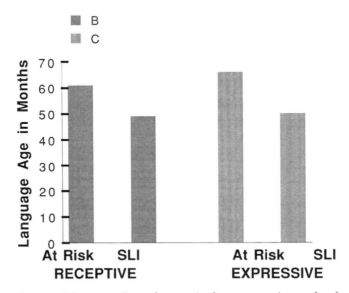

Figure 3   Mean receptive and expressive language age in months of
specific language-impaired (SLI) and at-risk children at 60 months.

tion of the comparative difficulty of the two tests. However, what is usually the case is that receptive abilities are far ahead of productive abilities.

The posited comprehension problem may be due to difficulties in identifying differences among phonological sequences and then relating these differences to objects and attributes of objects in the environment or, in other words, difficulties in only some part of the process, phonological or semantic, or pragmatic. Some studies have been carried out examining SLI and normally developing children's ability to learn novel words (Schwartz, 1991). These novel words were composed of phonological segments that were either a part of the child's repertoire or were not. It was found that normally developing children relied much more heavily on their available phonological repertoires to acquire new words than did the impaired children. This suggests that the part of the lexical acquisition process that involves comparison of new with old phonological representations may be impaired, perhaps a left-hemisphere process. There are, however, language-disordered children who have both a limited lexicon and a limited phonological repertoire (Stoel-Gammon, 1991), perhaps involving both hemispheres.

The comparisons made above to left- and right-lesioned children are, to put it mildly, highly speculative, although theoretically possible. We simply do not know at this time what underlying damage may exist in the brains of SLI children. Given new brain-imaging techniques, it is possible that we will eventually know. For now, we can only examine the behavioral data that have been obtained and speculate about underlying causes. This behavior suggests that there are differences among SLI children. It is not clear whether these differences are related to underlying nervous system differences or not.

A great many studies of both reading impaired and dyslexic children have examined their confrontation-naming abilities (for example, Denckla & Rudel, 1976; Wolf, 1984). Significant differences between dyslexic, reading impaired, and normally developing children have been found in these abilities. Both differences in latencies and in error scores occur. Some researchers have suggested that these errors are similar to those made by left-hemisphere-damaged adults after some recovery (Denckla & Rudel, 1976). There have been somewhat similar findings with oral language-impaired children. Depending on the age of the subjects, both lack of knowledge of lexical items and retrieval difficulties have been observed. For example, younger SLI children may have difficulty in naming letters, numbers, colors, or objects because they have yet to acquire the names for these items. They then may have difficulty in retrieving the names as rapidly as their age-matched peers. These behaviors are quite similar to those found in the previously cited study of brain-lesioned children (Aram et al., 1987).

A study (Fried-Oken, 1984, 1987) was carried out to examine the confrontation-naming skills of a group of language-impaired children when the set of stimulus pictures was presented twice. It was hypothesized that this might parcel out lack of knowledge of a name from word retrieval problems. Thirty children aged 4 to 9 years with measured language impairment were matched with 30 children who were developing language normally. The 50 items developed for the experiment could be divided into five semantic categories. The items represented words that were presumably learned early and included half high-frequency and half low-frequency words in each category. The items were representative of objects that were operative, palpable, and sensorially vivid. Between the first and second presentation of the task the experimenter provided the names of items that the children could not identify. Response latencies were measured and responses were divided into various error categories. The findings indicated that language-impaired children took significantly longer to respond to the items. There was no effect of semantic category on naming rate but there was a frequency effect, and, finally, a great deal of the latency effect could be accounted for by lack of knowledge of a name. The language-impaired children made a significantly greater number of errors, and there was an age-by-error interaction with the older language-impaired children making fewer errors than the younger but continuing to make more errors than the younger, normally developing children. Finally, there were significant differences between the groups in the types of errors made after eliminating the responses that indicated lack of knowledge of the name. These remaining responses could then be considered retrieval errors. The normally developing children most frequently made errors that were related both semantically and perceptually to the target item. The language-impaired children, although also frequently making these types of errors, also made much more frequent circumlocutory, purely semantic and nonrelated errors. These results are shown in Table 2. These differences in error types for the two groups might be indicative of differences in retrieval strategies. It is not clear whether they can be characterized as either left- or right-hemisphere processing problems.

In addition to confrontation naming studies, difficulties of language-impaired children have been examined under other conditions similar to those used with lesioned children. German (1982) in an experiment using picture naming, sentence completion, and naming to description found that 7–12-year-old language-impaired children produced a significantly greater number of "I don't knows," reformulations, and word substitutions than did normal controls. The substitutions were based on semantic or visual similarity to the target word or began with the same sound. In examining the word-finding difficulties of these children in a sample of spon-

Table 2 Percentages of Types of Errors in Language-Impaired Children's Responses on a Confrontation Naming Task[a]

Error types	Normal (%) Ages (years)			Language-impaired (%) Ages (years)		
	4	6	8	4	6	8
Phonological	6.5	0.0	7.3	9.6	11.4	8.5
Perceptual	7.3	2.6	1.8	8.2	1.4	2.8
Semantic	4.9	13.2	12.7	12.3	20.0	21.1
Ordinate	9.8	7.9	0.0	4.1	5.7	1.4
Circumlocution	7.3	0.0	0.0	11.0	11.4	15.5
Unrelated	12.2	0.0	1.8	21.9	17.1	15.5
Semantic/Perception	53.7	65.8	72.7	28.8	31.4	31.0
Semantic/Perception/Phonology	0.0	2.6	3.6	4.1	1.4	4.2

[a]Data adapted from Fried-Oken, 1984.

taneous speech, German (1987) found that the normally developing children produced longer sentences and a greater total number of sentences, and that there were a significantly greater number of reformulations and substitutions of words in the spontaneous speech of the impaired children. Thus, both in naming and in discourse experiment conditions the children exhibited similar behavior. These findings were replicated to some extent in another study (German & Simon, 1991), analyzing narrative productions of children in grades 1 to 6. Overall, the language-impaired children were not less productive than the normally developing children, as measured by number and length of T-units, as was found in the previous study. This was, perhaps, an effect of the genre of the discourse. However, they produced significantly more word substitutions, reformulations, repetitions, insertions, and empty words such as "this," or "that thing." In summary, research results indicate the prominence of word-finding problems among SLI children as well as word-acquisition problems. These problems are similar, to some extent, to the word-acquisition and word-finding problems of lesioned children.

SOME TENTATIVE CONCLUSIONS

In each of the areas discussed in this chapter, lexical development of normally developing children, word acquisition, and word finding in lesioned children and in children with suspected underlying brain damage, there are many questions that remain. There are questions about what word knowledge is stored and how it is retrieved and where it is stored and retrieved from. Most importantly, the effect of the content and structure of the word and the processing required dependent on this structure and content needs to be determined. The answers that have been obtained suggest that lexical acquisition and retrieval involves many processes. It seems safe to say that some of these processes appear to involve different parts of the left and right hemisphere, or, in some instances, both hemispheres, depending on age at time of lesion. It also seems safe to say that the problems observed in lesioned and suspect children are both similar to and different from the problems observed in brain-damaged adults. The differences observed may be due to developmental changes that occur both in the organization of the nervous system and in cognitive and social maturation of the child. This maturation over time makes the problem of understanding the relation between word acquisition and word finding and nervous system functioning in children a very complex problem but, nevertheless, an intriguing and important one. If one could better understand these relations then intervention techniques that are more appropriate could be planned and carried out.

ACKNOWLEDGMENTS

This chapter was written while the author was supported in part by National Institute of Deafness and Other Communication Disorders (NIDCD) grant DC00537. The author is grateful to Dorothy Aram for making her aware of recent research on lexical acquisition and naming in lesioned children.

REFERENCES

Anglin, J. (1993). Vocabulary development: A morphological analysis. *Monographs of the Society for Research in Child Development*, Serial no. 238, *58*, No. 10.

Aram, D., Ekelman, B., Rose, D., & Whitaker, H. (1985). Verbal and cognitive sequelae following unilateral lesions acquired in early childhood. *Journal of Clinical and Experimental Neuropsychology, 7*, 55–78.

Aram, D., Ekelman, B., & Whitaker, H. (1987). Lexical retrieval in left and right brain lesioned children. *Brain and Language, 31*, 61–87.

Bates, E., O'Connell, B., & Shore, C. (1987). Language and communication in infancy. In J. Osofsky (Ed.), *Handbook of infant development 2nd ed.* (pp. 149–203). New York: John Wiley.

Bates, E., & Thal, D. (1991). Associations and dissociations in child language development. In J. Miller (Ed.), *Research on child language disorders.* (pp. 147–168). Austin, TX: Pro-Ed.

Bishop, D. Language development after focal brain damage. In D. Bishop & K. Mogford (Eds.), *Language development in exceptional circumstances* (pp. 203–219). Hillsdale, NJ: Lawrence Erlbaum Associates.

Bloom, L. (1991). *Language development from two to three.* New York: Cambridge University Press.

Bullock, D., Liederman, J., & Todorovic, D. (1987). Reconciling stable asymmetry with recovery of function: An adaptive systems perspective on functional plasticity. *Child Development, 58,* 689–712.

Carey, S. (1978). The child as word learner. In M. Halle, G. Miller, & J. Bresnan (Eds.), *Linguistic theory and psycholinguistic reality* (pp. 264–293). Cambridge, MA: M.I.T. Press.

Clark, E. (1983). Meanings and concepts. In J. Flavell & E. Markman (Eds.), *Handbook of child psychology, Vol. III* (pp. 787–840). New York: John Wiley.

Curtis, S., Katz, W., & Tallal, P. (1992). Delay versus deviance in the language acquisition of language impaired children. *Journal of Speech and Hearing Research, 35,* 373–383.

Dale, P., Bates, E., Reznick, S., & Morisset, C. (1989). The validity of a parent report instrument of child language at 20 months. *Journal of Child Language, 16,* 239–250.

Denckla, M., & Rudel, R. (1976). Rapid "automatized" naming (R.A.N.): Dyslexia differentiated from other learning disabilities. *Neuropsychologia, 14,* 471–479.

Dennis, M. (1980a). Capacity and strategy for syntactic comprehension after left or right hemidecortication. *Brain and Language, 10,* 287–307.

Dennis, M. (1980b). Language acquisition in a single hemisphere: Semantic organization. In D. Caplan (Ed.), *Biological studies of mental processes,* (pp. 404–438). Cambridge, MA: M.I.T. Press.

Dromi, E., Leonard, L., Shleitman, M. (1993). The grammatical morphology of Hebrew speaking children with specific language impairment: Some competing hypotheses. *Journal of Speech and Hearing Research, 36,* 760–771.

Eisele, J., & Aram, D. (1993). Differential effects of early hemisphere damage on lexical comprehension and production. *Aphasiology, 7,* 513–523.

Feldman, H., Holland, A., Kemp, S., & Janosky, J. (1992). Language development after unilateral brain injury. *Brain and Language, 42,* 89–102.

Fried-Oken, M. (1984). *The development of naming skills in normal and language deficient children.* Unpublished doctoral dissertation, Boston University.

Fried-Oken, M. (1987). Qualitative examination of children's naming skills through test adaptation. *Language, Speech and Hearing Services in the Public Schools, 18,* 206–216.

Fromkin, V. (1993). Speech production. In J. Berko-Gleason & N. Bernstein-Ratner (Eds.), *Psycholinguistics* (pp. 271–300). Fort Worth: Harcourt Brace Jovanovich.

German, D. (1982). Word finding substitutions in children with learning disabilities. *Language, Speech and Hearing Services in the Public Schools, 13,* 223–230.

German, D. (1987). Spontaneous language profiles of children with word-finding problems. *Language, Speech and Hearing Services in the Public Schools, 18,* 217–230.

German, D., & Simon, E. (1991). Analysis of word-finding skills in discourse. *Journal of Speech and Hearing Research, 34,* 309–336.

Gibson, J., & Spelke, E. (1983). The development of perception. In J. Flavell & E. Markman (Eds.), *Handbook of child psychology, Vol. III* (pp. 1–76). New York: John Wiley.

Hirsh-Pasek, K., Reeves, L., & Golinkoff, R. (1993). Words and meanings: From primitives to complex organization. In J. Berko-Gleason & N. Bernstein-Ratner (Eds.), *Psycholinguistics* (pp. 134–199). Fort Worth: Harcourt Brace Jovanovich.

Kelly, D., & Rice, M. (1994). Preferences for verb interpretation in children with specific language impairment. *Journal of Speech and Hearing Research, 37,* 182–192.

Kiessling, L., Denckla, M., & Carlton, M. (1983). Evidence for differential hemispheric function in children with hemiplegic cerebral palsy. *Developmental Medicine and Child Neurology, 25,* 727–734.

Kuhl, P. (1990). Toward a new theory of the development of speech perception. In S. Furui (Ed.), *Proceeding of the international conference on spoken language processing* (Vol 2, pp. 745–748). Tokyo: Acoustical Society of Japan.

Lenneberg, E. (1967). *The biological foundations of language.* New York: John Wiley & Sons.

Leonard, L. (1991). The cross-linguistic study of language impaired children. In J. Miller (Ed.), *Research in child language disorders: A decade of progress* (pp. 379–386). Austin, TX: Pro-Ed.

Markman, E. (1991). The whole- object, taxonomic and mutual exclusivity assumptions as initial constraints on word meaning. In S. Gelman & J. Byrnes (Eds.), *Perspectives on language and thought: Interrelations in development* (pp. 72–106). New York: Cambridge University Press.

Marslen-Wilson, W., & Tyler, L. (1980). The temporal structure of spoken language understanding. *Cognition, 8,* 1–71.

Medin, D., & Schaffer, M. (1978). Context theory of classification learning. *Psychological Review, 85,* 207–238.

Menyuk, P. (1988). *Language development: Knowledge and use.* New York: Harper Collins.

Menyuk, P. (1992). Early communicative and language behavior. In J. Rosenblith (Ed. and Author) *In the beginning: Development from conception to age two.* (2nd ed.) (pp. 428–455). Newbury Park, CA: Sage Publications.

Menyuk, P. (1993). Children with specific language impairment (developmental dysphasia): Linguistic aspects. In G. Blanken, J. Dittmann, H. Grimm, J. Marshall, & C. W. Wallesch (Eds.), *Linguistic disorders and pathologies: An international handbook* (pp. 606–625). Berlin: Walter de Gruyter.

Menyuk, P., Liebergott, L., Chesnick, M., Korngold, B., D'Agostino, R., & Belanger, A. (1991). Predicting reading problems in at risk children. *Journal of Speech and Hearing Research, 34,* 893–901.

Menyuk, P., Liebergott, J., & Schultz, M. (1995). *Language developments in infants: Premature and full-term.* Hillsdale, NJ: Erlbaum.

Morgan, J., & Saffran, J. (1995). Emerging integration of sequential and suprasegmental information in preverbal speech segmentation. *Child Development, 66,* 911–936.

Oetting, J., Rice, M., & Swank, L. (1995). Quick incidental learning (QUIL) of words by school-aged children with and without SLI. *Journal of Speech and Hearing Research, 38,* 434–445.

Paul, R., & Alford, S. (1993). Grammatical morpheme acquisition in 4-year-olds with normal, impaired and late-developing language. *Journal of Speech and Hearing Research, 36.* 1993.

Pinker, S. (1984). *Language and learnability.* Cambridge, MA: M.I.T. Press.

Pinker, S. (1987). The bootstrapping problem in language acquisition. In B. MacWhinney (Ed.), *Mechanisms of language acquisition* (pp. 73–193). Hillsdale, NJ: Erlbaum.

Quinn, P. (1994). The categorization of above and below spatial relations by young infants. *Child Development, 65,* 58–69.

Rescorla, L., & Goossens, M. (1992). Symbolic play development in toddlers with expressive specific language impairment (SLI-E). *Journal of Speech and Hearing Research, 35,* 1290–1302.

Rescorla, L., & Schwartz, E. (1990). Outcome of toddlers with specific expressive language delay. *Applied Psycholinguistics, 11,* 393–407.

Rice, M., Buhr, J., & Nemeth, M. (1992). Fast mapping word-learning abilities of language-delayed preschoolers. *Journal of Speech and Hearing Disorders, 35,* 33–42.

Robinson, R. (1981). Equal recovery in child and adult brain? *Developmental Medicine and Child Neurology, 23,* 379–382.

Rosch, E. (1975). Cognitive representations of semantic categories. *Journal of Experimental Psychology: General, 104,* 192–233.

Schwartz, R. (1991). Lexical acquisition and processing in specific language impairment. In J. Miller (Ed.), *Research on child language disorders: A decade of progress* (pp. 339–348). Austin, TX: Pro-Ed.

Siegler, R. (1991). *Children's thinking* (2nd ed.). Englewood Cliffs, NJ: Prentice-Hall, Inc.

Stoel-Gammon, C. (1991). Issues in phonological development and disorders. In J. Miller (Ed.), *Research on child language disorders: A decade of progress* (pp. 255–266). Austin, TX: Pro-Ed.

Thal, D., Marchman, V., Stiles, J., Aram, D., Trauner, D., Nass, R., & Bates, E. (1991). Early lexical development in children with focal brain injury. *Brain and Language, 40,* 491–527.

Thomas, D., Campos, J., Shucard, D., Ramsey, D., & Shucard, J. (1981). Semantic comprehension in infancy: A signal detection analysis. *Child Development, 52,* 798–803.

Tomblin, B., Freese, P., & Records, N. (1992). Diagnosing specific language impairment in adults for the purpose of pedigree analysis. *Journal of Speech and Hearing Research, 35,* 832–843.

Vargha-Khadem, F., O'Gorman, A., & Watters, G. (1985). Aphasia in children with "prenatal," vs. postnatal left hemisphere lesions: A clinical and CT scan study. *Brain, 108,* 677–696.

Vygotsky, L. (1962). *Thought and language.* Cambridge, MA: M.I.T. Press.

Watkins, R., & Rice, M. (1991). Verb particle and preposition acquisition in language-impaired preschoolers. *Journal of Speech and Hearing Research, 34,* 1136–1141.

Waxman, S. (1991). Convergences between semantic and conceptual organization in the preschool years. In S. Gelman & J. Byrnes (Eds.), *Perspectives on language and thought: Interrelations in development* (pp. 107–145). New York: Cambridge University Press.

Wittelson, S. (1977). Early hemisphere localization and interhemispheric plasticity. In S. Segalowitz & F. Gruber (Eds.), *Language development and neurological theory* (pp. 213–287). New York: Academic Press.

Wittelson, S. (1987). Neurobiological aspects of language in children. *Child Development, 58,* 653–688.

Wolf, M. (1984). Naming, reading and the dyslexias. *Annals of Dyslexia, 34,* 87–115.

Woods, B., & Carey, S. (1979). Language deficits after apparent clinical recovery. *Annals of Neurology, 6,* 405–409.

Chapter 7

Naming in Normal Aging and Dementia of the Alzheimer's Type

Marjorie Nicholas, Christine Barth, Loraine K. Obler, Rhoda Au, and Martin L. Albert

INTRODUCTION

Elderly people often complain of an inability to remember names of people, places, and things. A disturbance in naming is also one of the many cognitive impairments demonstrated by individuals with dementia of the Alzheimer's type (AD). In the first section of this chapter we explore the phenomenon of age-related difficulties in finding the right word at the right time. We shall document changes in the ability to retrieve words linked to processes of normal aging, and examine how these naming problems affect everyday communication. We shall place observed language behaviors in the context of contemporary models of naming and attempt to analyze psycholinguistic aspects of naming in normal aging.

In Part II, we focus on the changes in naming ability associated with AD. In particular, we discuss research implicating a disturbance of semantic memory as the underlying mechanism of the naming impairment in subjects with AD. This section focuses on investigations of naming in AD with respect to naming errors, response to cues, and consistency of naming responses. We also discuss the relationship between dementia severity and naming disturbance in AD.

NAMING IN NORMAL AGING

Object Naming

The ability to produce a name for a given person, object, place, or action is commonly measured in a confrontation-naming task via a picture-nam-

ANOMIA: Neuroanatomical and Cognitive Correlates

ing paradigm. In this task, subjects are presented pictures of the to-be-named items. Cognitive operations involved in naming a picture include visuoperceptual processes (seeing and recognizing the picture), semantic processes (assessing the semantic or conceptual information of the item), lexical processes (retrieving the phonological and/or orthographic form of the word), and articulatory processes (saying the word). To investigate these processes, researchers measure the number of items named, as well as other qualities of naming ability, such as the time it takes to produce the name and the number and types of errors.

The most commonly used measure of confrontation naming is the Boston Naming Test (BNT) (Kaplan, Goodglass, & Weintraub, 1976, 1983). This test consists of a set of line drawings of 60 common objects (85 items are in the experimental version of the test) with items ranging in difficulty from bed or tree to abacus or trellis. The subject's task is to provide precise names for the pictured objects. Standardized sets of cues are given if the subject is not successful in producing a name. Cues are of two types: (1) phonemic, where the subject is given the initial sound of the word, and (2) semantic, where the subject is given information about the object such as its category membership ("it's an animal") or function ("it's used on dogs").

An early study of confrontation naming in healthy subjects using the BNT indicated a quantitative decline in naming ability with increasing age; older adults named fewer pictures than younger adults (Borod, Goodglass, & Kaplan, 1980). This empirical support for naming problems in the elderly in conjunction with common anecdotal reports of word-finding difficulty with increasing age sparked interest in research on naming changes with age.

At about this time, researchers in our own laboratory began a study that would provide additional information about language changes in normal aging and which could, by virtue of the longitudinal method of study, determine if age differences were caused by cohort effects. We tested healthy male and female adults, ranging in age from 30 to 79 and grouped into four age ranges: 30–39, 50–59, 60–69, and 70–79. Cross-sectional analyses from our first testing session provided further support for Borod et al.'s (1980) finding of decreased naming ability with age (Nicholas, Obler, Albert, & Goodglass, 1985). BNT results demonstrated a significant difference in number of correct responses among age groups, with the 70s group scoring significantly lower than all other age groups. Additionally, analyses of response to cues showed that, although older age groups received significantly more phonemic cues, there was no difference between the age groups in the ability to utilize such cues. That is to say, all age groups were helped equally by the cues.

In addition to examining naming scores, we also analyzed the types of errors. We found that the most common errors on the BNT for all subjects were semantically related to the target word. Comments and circumlocutions were also common errors. Younger subjects produced proportionally more semantically related errors, more errors that were both semantically and perceptually related to the target, and more phonologically related errors than did older subjects. Older subjects produced proportionally more circumlocutions and augmented correct responses (i.e., responses with an appended modifier) than younger subjects (Nicholas et al., 1985).

A study by M. Albert and her colleagues (Albert, Heller, & Milberg, 1988) found slightly different changes in error types and response to cuing with increasing age. First, their study replicated our previous result of a sharp decline in BNT performance after age 70, with their subjects in their 70s spontaneously naming significantly fewer items on the BNT than did the younger age groups. Second, circumlocutions, semantically related responses, nominalizations (words describing the function of the target word), and perceptual errors all increased in frequency with age. Third, unlike our results, they found age group differences even after cues were administered.

The results of our longitudinal study on naming and aging provide support for an age-related naming problem that is not due primarily to cohort effects (Au et al., 1995). This study followed 53 subjects from our original four age groups across a 7-year time span. The longitudinal results indicated that all groups except the 30-year-olds showed significant declines in BNT naming scores over time. Additionally, we found evidence that only the oldest group (people in their 70s) benefited less from cues over time. This could indicate a greater retrieval difficulty due to a decrease in processing efficiency not seen in younger subjects followed over time. Alternatively, these findings could argue for a problem in word retrieval in elderly adults that is qualitatively different from that seen in younger adults, although this seems less likely.

Not all researchers have reported age-related changes in naming ability. However, even when age group differences do not reach statistical significance, older age groups generally demonstrate worse performance than younger age groups (Goulet, Ska, & Hahn, 1994). Van Gorp and colleagues (Van Gorp, Satz, Kiersch, & Henry, 1986) and L. Nicholas and colleagues (Nicholas, Brookshire, MacLennan, Schumacher, & Porrazzo, 1989) both reported nonsignificant correlations between BNT score and age. L. Nicholas et al. (1989) did not report scores by age groups, but Van Gorp et al. (1986) noted greater performance variability and lower cutoff scores for their older subjects. Similarly, although LaBarge, Edwards, and Knesevich

(1986) argued against changes in naming ability with age, they still found a decline in number correct within their 60- to 85-year-old subjects. Given that response latencies may be a more sensitive measure of ease of retrieval, it is interesting to note that age-related slowing in name retrieval can be demonstrated, even when there are no differences in percentage correct (Thomas, Fozard, & Waugh, 1977).

Some researchers have argued that age effects may really be due to educational differences across age cohorts (LaBarge et al., 1986; Van Gorp et al., 1986). Although it is common for older groups in the United States to be less well educated due to societal changes and specific events such as the Depression, research has generally concluded that education is not a significant factor in age-related declines in word-finding ability (see Albert et al., 1988; LaBarge et al., 1986). However, that is not to say that education level does not affect naming ability as several studies have shown that it does (Barth, Nicholas, Au, Obler, & Albert, 1996; Le Dorze & Durocher, 1992; L. Nicholas et al., 1989).

In summary, several studies using the BNT have found evidence for word-finding difficulty with increasing age, especially after age 70. Even when results have not been statistically significant, they have indicated both lower scores and a broader range of performance in elderly subjects. Our own longitudinal studies of naming have provided further support for an age-related decrease in confrontation-naming ability.

The Tip-of-the-Tongue Phenomenon

Evidence of changes in word-finding ability with age also comes from studies investigating the tip-of-the-tongue (TOT) phenomenon. TOT is said to occur when a familiar person or object cannot be named despite successful access to information about the target word. Studies have commonly reported a larger number of TOTs in older adults, using both informal diary-keeping methods (Burke, MacKay, Worthley, & Wade, 1991; Cohen & Faulkner, 1986) and more formal procedures with laboratory-induced TOTs (Burke et al., 1991). In addition, most studies have found that older subjects have less access to information about the TOT target word (phonological and/or lexical information) than younger subjects do (Burke et al., 1991; Cohen & Faulkner, 1986; Maylor, 1990).

Using a naturalistic diary study of word-finding problems, Burke et al. (1991) found a significant effect of age on the number of TOTs reported. Both older (mean age 71.0) and midage (mean age 38.7) adults reported significantly more TOTs than younger (mean age 19.4) adults. Proper name TOTs accounted for 69% of all TOTs in both midage and older adults as compared to 58% for younger subjects. Despite the fact that older adults reported equally high "feeling of knowing" for the TOT target word, they had less access to partial information and reported fewer alternates to the target word than did younger adults. These findings were further supported by Burke et al. (1991) who, using a laboratory method to induce TOTs, found that older subjects had a larger proportion of TOTs than younger subjects for object names, adjectives, verbs, and names of famous people.

Consistent with the results of Burke et al. (1991) is Maylor's finding of an increase in proper name TOTs with age (Maylor, 1990; Maylor & Valentine, 1992). Using a task involving recognition and naming of famous faces, she found an increase in the number of unresolved TOTs with age (Maylor, 1990). Unresolved TOTs are instances where the target word is not found during the course of the task. Burke et al. (1991) suggested that the problem of proper noun naming in older adults could result in part from lessened recency of use that is likely to occur with increasing age.

In summary, the TOT phenomenon occurs more often with increasing age, for several types of words, and in both natural and experimental procedures. The most common type of word for which TOTs are reported are proper nouns (Burke et al., 1991). Any theory that accounts for word-finding difficulty in aging should thus consider why proper names are often a specific source of word-finding difficulty. (See Semenza, chap. 5, this volume.)

Action Naming

Research efforts in our own laboratory have recently been focused on changes in the ability to retrieve verbs with age. In other studies, healthy elderly subjects have been tested for verb-naming skills but only as controls for subjects with aphasia (e.g., Kohn et al., 1989; Miceli, Silveri, Villa, & Caramazza, 1984) or as part of investigations of general word-finding ability (e.g., Burke et al., 1991). There have been no investigations designed to establish norms for verb-naming ability with age.

Because of our interest in verb-naming abilities, we created the Action Naming Test (ANT) (Obler & Albert, 1979). The ANT was designed to be similar to the BNT both in overall form and administration. It consists of 55 line drawings of actions which range in frequency from items such as sleeping or sitting to knighting or proposing. We first used the ANT in a cross-sectional design in order to measure changes in ability to retrieve verbs with age. Similar to the BNT results, ANT results indicated an age-related decline in the ability to give an accurate name to an action picture, with healthy people in their 70s performing significantly worse than people in their 30s, 50s, or 60s (Nicholas et al., 1985).

These findings have been further supported by results from our longitudinal study of action naming ability (Barth et al., 1996). Healthy men and women in their 30s, 50s, 60s, and 70s were given the ANT three times over a 7-year time span. Results indicated that only the 30-year-old group did not show a significant decline in verb-naming ability over the three testing sessions. Thus, our longitudinal results indicated that decline in verb naming begins certainly by the sixth decade and continues with age. All subjects made more errors and more types of errors across time. Elderly and young subjects performed equally well when given cues.

In summary, results from studies of verb naming ability in healthy aging indicate that retrieval of verbs is vulnerable to the aging process. Both cross-sectional and longitudinal studies have supported this finding. Furthermore, problems in verb-naming ability with increasing age seem to be similar to other word-finding difficulties commonly found in aging.

Action Naming versus Object Naming

Several recent investigations have indicated qualitative differences between noun and verb production in aphasic subjects (Kohn et al., 1989; Miceli et al., 1984; Miceli, Silveri, Nocentini, & Caramazza, 1988). Given the declines in both noun and verb naming that have been seen in normal aging, we were also interested in comparing noun and verb naming results from our longitudinal study. Such a post hoc comparison requires an indirect approach, but in this case has the distinction of using data from many (but not all) of the same subjects for both studies, thus minimizing confounding subject factors. A direct comparison is difficult because of the near impossibility of equating the noun and the verb tests on other parameters, such as difficulty and word frequency.

In our longitudinal studies of both the BNT and the ANT, naming declines were seen over a 7-year time span in all age groups except the youngest age group of people in their 30s. Additionally, the BNT analyses indicated that the oldest group of subjects benefited less from cues, whereas ANT results showed that all age groups were able to use cues equally effectively. In general, subjects of all ages were able to name a higher percentage of the items on the ANT than on the BNT. It is possible that verb naming may not be affected to the same extent as noun-naming ability in the aging process. Alternatively, the ANT may be an easier test than the BNT. Support for this possibility comes from data that show that the mean percent correct on the ANT is greater than the mean percent correct on the BNT for all age groups (M. Nicholas et al., 1985).

To further investigate possible processing and/or organizational differences between nouns and verbs, we compared error types on both the ANT and the BNT. We discovered significant age-related changes in more error types on the BNT than on the ANT. More specifically, we observed significant differences among the four age groups or across time within age groups on both the BNT and the ANT for semantically related errors, perceptually related errors, fragments, and perseverations. However, the BNT results showed significant changes as well in phonemically related errors, circumlocutions, and comments, whereas similar analyses on the ANT yielded no significant changes for any of these error types across age groups or across time within age groups. Thus there are subtle yet important differences found in the qualitative analysis of error types. Although it is difficult to interpret the meaning of the specific differences, the overall pattern of significant findings for these qualitative scores may indicate characteristic differences between noun and verb naming as these abilities interact with age.[1]

Our findings of qualitative differences between nouns and verbs, however, are consistent with related literature. Investigations of naming in both language-impaired and normal populations have indicated contrasts between noun and verb-naming abilities, although the findings have not been uniform (Kohn et al., 1989; Miceli et al., 1984, 1988). Generally, the normal subjects in these studies have performed near ceiling. Thus, the lack of clear consistent findings could relate to the difficulty level of the naming task. Also, because the normals tested in these studies functioned as controls for an aphasic population, only a small number (5–20) were tested. The lack of uniform findings within these normal subjects could thus relate to the heterogeneity associated with a small pool of elderly subjects.

A dissociation between noun- and verb-naming abilities has recently received anatomical support. Using PET scanning, Damasio & Damasio, (1992) showed that a region of the lateral and inferior dorsal frontal cortex is activated during verb production. A. Damasio and Tranel (1993) found a double dissociation between noun and verb retrieval using three pa-

---

[1]Although an indirect comparison of the BNT and the ANT cannot demonstrate statistically significant differences between noun and verb naming in normal aging, these results indicate that the two tasks appear to induce qualitatively different naming performances. Nonetheless, the limitations of this comparison cannot be overlooked. The illustrations are, of course, conceptually different (in that actions vs. objects are pictured), and they were created by different artists, although the two tests are similarly constructed of line drawings. The ANT has fewer items (55) than the experimental version of the BNT (85 items) that we have used in our studies. This difference might interact with fatigue effects. Differences in frequencies of the target words may also contribute to discrepant findings between the ANT and the BNT. Although the frequencies of the pictured words on the two tests are not statistically different if the most common accepted responses are used (i.e., "running" for the picture of running), the BNT target words are less frequent than the ANT target words if the general word class of each target is considered (i.e., "run" for running), using norms from Francis and Kucera (1982).

tients. Two of their patients had lesions in the left anterior and middle temporal lobes and showed deficits in proper and common noun retrieval but were unimpaired in verb retrieval. Conversely, the third patient, with a lesion in the left premotor cortex, had impaired verb retrieval with intact noun retrieval and no other linguistic impairments.

Finally, a study by Caramazza and Hillis (1991) using homonyms suggests that the dissociation between nouns and verbs is truly a result of grammatical class differences rather than just differences in the word forms of nouns and verbs. Homonyms, of course, are words that have identical orthographic and phonological forms but different meanings and, in the Caramazza and Hillis study, belong to different grammatical classes (i.e., a smile, to smile, or a knight, to knight). Caramazza and Hillis found support for a division within the lexicon based on grammatical class. They tested two patients, both of whom produced fewer verbs than nouns, and found that these differences were maintained with the use of homonyms.

The results of our indirect ANT and BNT comparison offer support for a grammatical class difference within the lexicon suggested by Caramazza and Hillis (1991) and Damasio and Tranel (1993). Although further work needs to be done to clarify specifically where and how the noun–verb dissociation exists, our results are consistent with a cognitive processing and/or anatomical dissociation between nouns and verbs in the language system.

Naming and Discourse

We have seen in our review of research on the BNT, the TOT phenomenon, and the ANT, that naming performance declines with age. Abundant evidence exists for word-finding difficulty with age using other measures as well. These other measures include tasks that elicit one-word responses such as naming to definition (Bowles & Poon, 1985) and tasks that measure word-finding ability in running speech (Cooper, 1990; Heller & Dobbs, 1993). Discourse studies have the advantage of looking at word-finding difficulty in a task most similar to spontaneous speech.

Cooper (1990) recently reported an increase in indefinite words and longer pauses among older subjects in an oral picture description discourse task. Heller and Dobbs (1993) found a larger number of "hedges" and incorrect object labels in older subjects in a video description discourse task. These differences in discourse production in young and older subjects may reflect difficulties with finding names with age. Our own results from a study of discourse production using the Cookie Theft picture from the Boston Diagnostic Aphasia Examination (Goodglass & Kaplan, 1983) also point to increased use of indefinite words such as "thing" with advancing age (see Obler, 1980; Sandson, Obler, & Albert, 1987).

In summary, there is evidence for a mild naming impairment in discourse production with increasing age. At the very least, elderly adults may find these word-retrieval problems a nuisance. However, as Kemper (1991) suggests, such problems could ultimately contribute to the "social isolation and withdrawal" that many older adults experience as they age.

Naming, Aging, and Cognitive Models

Because research results have generally converged on a finding of increasing word-retrieval difficulties with age, one must ask why this occurs. In attempting to answer this question, we may look at the instances where research on language and memory has intersected, particularly with respect to the role of semantic memory processes in word retrieval. Semantic memory (Tulving, 1983) refers to the permanent store of conceptual information that includes knowledge about word meanings. One possibility for why naming problems increase with age is that information about the items to be named is lost from semantic memory. In fact, this idea has been much debated in investigations of naming in Alzheimer's disease (see the second part of this chapter), but it has received little support from research in normal aging. Several studies have found evidence that semantic priming does not change with age (see Light & Burke, 1988; Madden, Pierce, & Allen, 1993). There is also evidence that semantic organization is similar in older and younger adults although older adults may be less consistent in their semantic associations (Brown & Mitchell, 1991; Light & Burke, 1988; Puglisi, Park, & Smith, 1987). Furthermore, the helpfulness of cuing on the BNT supports the idea that information is available but is not easily accessed (Nicholas et al., 1985). Although it is possible that there may be changes in the organization of semantic information with age (see Bowles, Obler, & Poon, 1989), the general findings support changes in word retrieval due to a lexical access problem rather than a change in the structure of semantic memory.

If the contents of semantic memory remain intact and instead lexical access begins to deteriorate, how may we explain this phenomenon? Investigators generally agree that one contributing factor may be cognitive slowing associated with aging, resulting in a decreased rate of information processing (Klatzky, 1988; Salthouse, 1988). Recent research has investigated a slowing factor specifically related to lexical information processing (Madden et al., 1993; Myerson, Ferraro, Hale, & Lima, 1992). However, it should be noted that only controlled and effortful lexical access, rather than automatic processes, seem to be affected by the aging process (Stern, Prather, Swinney, & Zurif, 1991). Using a continuous list priming paradigm, Stern and colleagues found that elderly subjects showed the same priming effects as younger subjects for words presented at short inter-stimulus intervals when only automatic processes are activated.

Besides a general slowing associated with aging, what makes giving names such a vulnerable process in the elderly? Several other factors related to processing changes have been proposed, including decreased peak activation levels (Klatzky, 1988), less available conscious search mechanisms (Bowles et al., 1989), decreased working memory resources (Light & Burke, 1988), and the Transmission Deficit Hypothesis (labeled by Burke et al., 1991). In the model of word-retrieval problems proposed in the Transmission Deficit Hypothesis, Burke et al. (1991) explain that "TOTs occur when the connections between lexical phonological nodes become weakened due to infrequent use, nonrecent use, and aging, causing a reduction in the transmission of priming" (p. 542). Priming here refers to the parallel spread of activity along the connections in a nodal network. The model of word-finding proposed in this theory of naming problems is consistent with other word-retrieval models proposed in the cognitive literature (see lexical hypothesis: Glaser, 1992; cascade model: Humphreys, Riddoch, & Quinlan, 1988).

The Transmission Deficit Hypothesis accounts for several findings related to naming difficulties and may explain why a reduction in priming transmission can coexist with intact semantic priming in the elderly. This paradox is accounted for by a network organization that allows for convergence of semantic information that can overcome priming deficits and allow for intact semantic priming. In contrast, convergence and summation of priming cannot occur with phonological nodes, leaving older adults with word-finding difficulty (Burke et al., 1991).

The Transmission Deficit Hypothesis may also account for the common problem of the age-related difficulty in retrieving proper nouns in particular. The vulnerability of proper names to the TOT state results from the lack of semantic connections with these particular names, resulting in fewer chances for the activation between nodes necessary for successful naming. With more abundant connections, convergence of priming is more likely to occur in the lexical system, thus raising the likelihood of activating related nodes and leading to successful word retrieval (Burke et al., 1991).

Differences in retrieval of nouns versus verbs may also be explained by the Transmission Deficit Hypothesis. Just as proper nouns are more sensitive to TOTs due to fewer connections in the lexicon, so too may nouns be more sensitive to word-finding difficulties with age than verbs. Nouns and verbs naturally function differently in language, and it may be reasonable to assume that they have different representations in the brain. Conceivably, nouns may be linked to a few associated nouns, whereas verbs may have a wider connection of associates from several different form classes

(Gentner, 1981). By the Transmission Deficit Hypothesis, one might expect the richer connections for verbs to increase the likelihood that semantic priming can summate on nodes to overcome lexical retrieval difficulties.

In summary, research has suggested that word-retrieval difficulty in aging may result from changes in lexical access capabilities rather than breakdown of the semantic network or other type of information loss. The Transmission Deficit Hypothesis, a recent explanation of the cause of word-finding problems, proposes that a reduction in the transmission of priming accounts for naming difficulties with age. This hypothesis includes a model of naming that accounts for several recent research findings related to naming and aging including the preservation of semantic priming despite word-finding problems, the particular difficulty with proper noun naming, and differences in the ability to name nouns versus verbs.

Summary

We have seen that word-retrieval problems are common with increasing age. This is true especially for proper and common nouns, but also for other word types such as verbs. Word-finding difficulty is apparent in several types of tasks ranging from naming tests such as the BNT and ANT to spontaneous discourse. Although the exact cause for this word-finding difficulty with age is still being debated, it is generally agreed that changes in processing underlie the naming decline. A current theory of word-finding difficulty that fits well with research results is the Transmission Deficit Hypothesis, which suggests that retrieval problems are a consequence of the decrease in priming transmission that occurs with increased age.

In the next section of the chapter we discuss word-finding problems in AD. Many of the same issues that have been investigated in naming in normal aging are also being looked at in studies of naming in AD. Just as in studies of normal aging, researchers have focused on contrasting theories of naming deficits that focus on disorders of access to lexical and semantic knowledge versus loss or deterioration of that knowledge.

NAMING IN DEMENTIA OF THE ALZHEIMER'S TYPE

Alzheimer's disease is one of the most common causes of dementia in older adults, resulting in a range of behavioral and cognitive impairments. Within the realm of language, naming difficulty has been observed repeatedly in patients with AD. Indeed, in some individuals, moderate to severe anomia may be one of the most striking characteristics of spontaneous discourse. Of the skills required for the production of meaningful appropriate spontaneous speech, the use of syntax and articulation are often preserved late into the course of the disease, but word-finding ability is frequently one of the first language skills to show impairment (Bayles & Kaszniak, 1987; Obler & Albert, 1984). We recognize that the diagnosis of Alzheimer's disease cannot be made with certainty without a post mortem study. References to AD in this chapter should be understood as references to probable Alzheimer's disease.

Recall the components of the task of confrontation naming mentioned above. Although the contributions of semantic and lexical processes to naming have been debated in the literature on normal aging, in AD the first component, visuoperceptual processes, has also been called into question. Some researchers have suggested that visuoperceptual factors account for some, and in some cases a significant portion, of the naming impairment in patients with AD (Cormier, Margison, & Fisk, 1991; Kirshner, Webb, & Kelly, 1984; Shuttleworth & Huber, 1988). Because patients with AD had difficulty naming visually degraded stimuli, and because some of their misnamings were names of items that were visually similar items to the target, it was suggested that at least some of their naming difficulty was due to visuoperceptual factors. However, visual manipulation of the target affected the performance of normal controls as well, often to a similar degree. In other words, although visuoperceptual factors could affect naming in patients with AD if the target were degraded or otherwise affected, these factors often had a similar effect on normal subjects. (For further discussion of visual deficits in AD, see Cronin-Golomb, Corkin, & Rizzo, 1991.)

In recent years, evidence has been converging to implicate the semantic and/or lexical system as the primary source of the naming impairment in AD, rather than the visuoperceptual system. Some researchers have suggested that a central problem in the organization or storage of concepts in semantic memory underlies not only the naming difficulty observed in AD patients, but also a number of other language behaviors observed in other tasks. However, Nebes and colleagues (Nebes, 1989; Nebes & Brady, 1988, 1990; Nebes, Martin, & Horn, 1984) have argued that semantic information is intact in AD but that conscious access to that information may be impaired. This argument is similar to accounts of the naming impairment in normal aging that were discussed above. Let us first consider some of the information implicating a disturbance in semantic memory.

A Disturbance in Semantic Memory

Information pointing to a disturbance in semantic memory as the underlying mechanism of the naming impairment in AD has come from several sources. Among these are analyses of error response types, responsivity to semantic and phonological cues, response consistency across time, and response consistency across tasks. Each of these topics is reviewed briefly below.

Analyses of error responses

A number of studies have investigated both the quantity and the type of errors produced in naming tasks in order to further characterize the naming impairments of patients with AD (Bayles, Tomoeda, & Trosset, 1990; Bowles, Obler, & Albert, 1987; Goldstein, Green, Presley, & Green, 1992; Kirshner et al., 1984; LaBarge, Balota, Storandt, & Smith, 1992; Martin & Fedio, 1983; Smith, Murdoch, & Chenery, 1989; Shuttleworth & Huber, 1988). When a patient correctly names a picture we assume that both semantic and lexical information were accessed for that item. In contrast, when a patient makes no response to an item, it is not known whether any semantic or lexical information was retrieved. Many responses, however, fall in between these two extremes; and it is to these responses that researchers turn in order to search for clues about the semantic and/or lexical processes involved in naming impairments. For example, when asked to name a picture of a unicorn, if a patient says, "that mythical animal, like a horse with the horn," we can assume that the picture was correctly perceived and that at least some of the appropriate semantic conceptual information was available to the patient, even though the lexical label "unicorn" may have been inaccessible. In contrast, if a patient says "deer," it is less clear that the picture was correctly perceived or that the correct semantic information was available.

Many studies analyzing naming errors in both normal aging and AD have relied on qualitative error coding schemes that group responses into categories such as "perceptual error," "semantic error," and so on. Analyses of this type have suggested that patients with AD are more likely than normal elderly subjects to produce errors that are unrelated to the target (Bowles et al., 1987; LaBarge et al., 1992) or errors that are semantic associates of the target (Bowles et al., 1987; Goldstein et al., 1992). In general, results have indicated that naming errors due to misperceptions of the target were comparatively rare in AD (Bayles & Tomoeda, 1983; LaBarge et al., 1992; Martin & Fedio, 1983, Kirshner et al., 1984; Smith et al., 1989).

Error categorization schemes can be difficult to use because responses do not always fit into any single category, and reliable, operational definitions of the categories are difficult to develop. For example, in some studies, "semantic" responses are coded separately from "circumlocutions" even though they may both provide semantic information. Bowles et al. (1987), for example, divided errors into (1) near synonyms and adequate circumlocutions, (2) semantically related, and (3) semantically unrelated responses. Both of the first two response types provide some semantic information relative to the target. In another study, these may have been considered together as "semantically related." Moreover, some "perceptual" responses, for example, "sticks" for the target *stilts,* may also be semantically related to the target. Placing errors of this type in a category marked "perceptual" could be misleading because the semantic relationship is ignored.

In a recent study conducted in our laboratory (Nicholas, Obler, Au, & Albert, 1996) we avoided the use of an a priori error categorization method by rating *all* first-error responses made by a group of 23 patients with AD on the BNT (Kaplan, Goodglass, & Weintraub, 1983) on a five-point scale of semantic relatedness (1 = not at all similar in meaning; 5 = very similar in meaning). Patients were also subdivided into two groups based on severity of dementia in order to investigate the relationship between dementia severity and naming performance (see later section in this chapter on severity). Given the semantic deficit hypothesis, we predicted that the errors of patients with AD would be rated lower in semantic relatedness to the target than the errors of normal elderly controls. Surprisingly, this was not the case. Although all AD subjects were considerably worse namers, error responses of subjects with either mild or moderate AD were rated as semantically related to the targets as the errors of the controls were.

Analyses of naming errors necessarily have a somewhat limited application to the question of whether or not there is a central semantic system disruption underlying the naming impairment in AD. There may be other factors that lead to the production of naming errors that are unrelated to a central semantic deficit. Given these limitations, however, our results do not support the notion that progressive deterioration of semantic information in AD is reflected in naming errors.

Response to cues

Many popular tests of visual confrontation naming ability such as the BNT provide for the presentation of cues if a subject experiences difficulty in naming. On the BNT, if a subject makes no response or produces a name indicative of a misperception of the target, the subject is given a semantic cue such as "used for cutting" for the target "scissors." If the subject is still unable to name the item, the phonemic cue is provided, usually the first two phonemes of the word. Ability to profit from a phonemic cue is often interpreted as an indication that the subject had the lexical label for the item on the "tip of the tongue." The phonemic cue provided enough information to enable a subject to access the entire lexical label. It is presumed that access to semantic information was not a problem. In fact, it is just this pattern of performance that normal elderly people show on naming tests, and their ability to respond to phonemic cues has been interpreted in this way.

In recent years, this interpretation of the response to phonemic cues has been questioned. Wingfield, Goodglass, and Smith (1990) have suggested an alternative explanation based on the phenomenon of completion, that is, the ability to use phonological cues to say a word without necessarily accessing the semantics of the word. In their study, both normal and aphasic subjects were able to complete words after hearing only a small portion of the initial sounds of the word, even without a picture present. To our knowledge this phenomenon has not yet been demonstrated in subjects with AD. But in an earlier study from our laboratory (Obler & Albert, 1984) we observed that subjects with AD were much more likely than normal elderly subjects to produce phonologically related but semantically unrelated words when given a phonemic cue on the BNT. That is, they could use the cue to access a phonological word form, but they were not always accurate at monitoring their output to ensure that the word that came to threshold was appropriate as a name for the pictured item.

In patients with AD, Neils, Brennan, Cole, Boller, and Gerdeman (1988) found a relationship between the ability to respond to phonemic cues and dementia severity. Only mildly demented subjects were able to benefit from phonemic cues, showing the pattern typical of normal elderly patients. Subjects with more severe dementia were unable to profit from cues, suggesting to the authors that the source of the naming failure was not simply a failure of lexical access. However, Funnell and Hodges (1991) presented a case report of a patient with AD who was followed for 2 years. Because this subject could comprehend 75% of the items she could not name, her anomia was attributed to impaired access to phonological word forms, rather than to a semantic deficit. This particular deficit became worse over time and phonemic cues became less and less effective. Once an item required phonemic cuing it was unlikely ever to be spontaneously named again. This finding was counter to the prevailing notion that failures of access are associated with inconsistent responding. We will consider this issue in the next section.

Consistency of naming performance

Consistency across time.   If semantic information about an item has deteriorated or become inaccessible in semantic memory, then an individual should show a consistent inability to name that item. On the other hand, if a patient shows an inconsistent ability to name, missing the name on one occasion yet retrieving it on another, we assume that the semantic information is intact, but lexical access mechanisms are faulty. With this reasoning, the consistency of naming performance has been the focus of several studies involving patients with AD. For example, Henderson, Mack, Freed, Kempler, and Andersen (1990) found that 81% of BNT items were either consistently named or not named by subjects with AD who were tested twice at 6 months apart. Similarly, Chertkow and Bub (1990) reported that their group of 10 AD patients failed to name 92.5% of the same items they failed initially when tested 1 month later. The high degree of consistency in both these studies was interpreted as indicative of a loss of information in semantic memory for those items that were consistently not named. Huff, Mack, Mahlmann, and Greenberg (1988) found a similar consistency effect for their subjects with AD. But they also found naming was consistent across time in a group of aphasics, suggesting a similar loss of information for this group.

The performance of the subject described by Funnell and Hodges (1991), as well as the performance of the aphasic group in the Huff et al. (1988) study, however, suggest perhaps that the interpretation of consistency of naming performance across time needs to be re-evaluated. The case reported by Funnell and Hodges had consistent naming impairment, yet did not have loss of information in semantic memory as evidenced by performance on other tests besides naming. This pattern of performance raises the possibility that a lexical label could be consistently irretrievable while semantic memory remained intact. Perhaps looking at consistency of performance relating to a given concept across tasks would speak to this question more convincingly.

Consistency across tasks. If semantic information for a specific concept were lost or profoundly disrupted we might expect that a variety of cognitive tasks requiring access to that information would all be disrupted as well. For example, if the concept of "pelican" were no longer part of semantic memory, a subject should not be able to name a picture of a pelican, answer questions about what a pelican is, or be able to point to a picture of a pelican upon hearing its name. Research into this particular issue with AD subjects has produced mixed results. Chertkow and colleagues (Chertkow & Bub, 1990; Chertkow, Bub, & Seidenberg, 1989) found error correspondences across tasks for specific items, suggesting "a loss of knowledge occurring within an amodal semantic component" (p. 432). Huff, Corkin, and Growdon (1986) found item consistency across naming and discrimination tasks, again suggesting a loss of semantic information. Similarly, Hodges, Salmon, and Butters (1992) found about 78% consistency of impairment on selected items across tasks. Generally, these studies have compared performance across a small number of tasks without paying particular attention to the variable of task difficulty.

Bayles, Tomoeda, Kaszniak, and Trosset (1991), by contrast, investigated this issue directly. They studied response consistency to 13 concepts presented in 11 different tasks in a group of 69 patients with AD. A subset of patients was also followed longitudinally. In contrast to the studies just discussed, their results indicated that few subjects showed consistency of impairment across tasks. These authors suggested that task difficulty played a major role in whether or not impairments were observed on individual items. Studies limiting the comparison to two tasks, for example, may have concluded that knowledge was lost from semantic memory, when really the tasks may have been too difficult for the patient to perform. In their battery, subjects who had missed an item earlier sometimes responded appropriately to that item when tested later with an easier test. They concluded that "answering the question of whether AD results in concept-specific loss from SM (semantic memory) may be impossible using tasks that require conscious processing by the AD patients" (p. 180).

The findings of this study by Bayles and colleagues suggest that evaluating consistency of performance is not always a straightforward endeavor. Their results suggest that task difficulty and dementia severity are variables that need to be carefully controlled in order to investigate the effects of dementia on semantic memory. It is intriguing that not one single patient of their sample of 69 patients showed convincing evidence of a concept-specific loss of information from semantic memory. After reviewing the literature on semantic memory impairments in AD, Nebes (1989) similarly concluded "that the magnitude and nature of the semantic deficit . . . is heavily influenced by stimulus and task variables" (p. 390).

Naming and Dementia Severity

We have seen in the previous sections that severity of dementia is an important factor in the evaluation of naming in AD. There is abundant evidence to suggest that the degree of semantic system involvement in naming performance may be related to severity of dementia in Alzheimer's disease (Huff et al., 1986; Shuttleworth & Huber, 1988). Generally, researchers have interpreted the naming errors of more severely demented subjects as indicative of deterioration of the semantic network (Goldstein et al., 1992). Mildly demented subjects, in contrast, sometimes behave similarly to normal elderly on naming tasks (LaBarge et al., 1992; Neils et al., 1988). For example, LaBarge et al. (1992) found that linguistically related errors were common early in the course of AD, but that no-content errors (errors in which no interpretation of the picture was evident, such as saying "I know that, but I can't think of the name") were more common with increasing severity of dementia. In this study, no-content errors were interpreted to reflect a primary disruption of semantic information. Comparing across tasks, these authors found that both normal elderly subjects and very mild AD patients could always pick the correct word on a multiple-choice task even if they failed to name it on the naming task. This pattern of performance was interpreted as reflective of a lexical access problem in both groups with no primary involvement of semantic memory. In contrast, patients with more advanced AD were able to recognize only 64%

of items on the multiple-choice tasks, reflecting increasing involvement of the semantic system as the disease progressed.

Another guiding concept in the literature on semantic memory impairments in AD suggests a progression of loss within semantic memory that begins with loss or disruption of the knowledge of attributes of concepts and then progresses to loss of categorical information (Abeysingh, Bayles, & Trosset, 1990; Grober, Buschke, Kawas, & Fuld, 1985; Hodges et al., 1992; Martin & Fedio, 1983; Schwartz, Marin, & Saffran, 1979; Warrington, 1975). This idea was partly based on error analysis studies that had reported a preponderance of categorical names as errors (e.g., "animal" for beaver). Bayles, Tomoeda, and Trosset (1990) compared naming performance to two other tasks: category recall and category recognition. Bayles et al. (1990) found that misnaming by responding with a superordinate was *not* commonly observed in AD. They also reported that when they controlled for task difficulty, the relative difficulty of category performance (recall or recognition) to naming performance increased with increasing severity of dementia.

Summary

Studies of naming errors, responsivity to cues, and response consistency across time and task have for the most part suggested that an impairment in semantic memory underlies at least part of the naming impairment of AD. Furthermore, it is clear that this impairment in semantic memory interacts with dementia severity. However, we have seen that it is still open to question whether the contents of semantic memory itself are deteriorating or whether access to that system becomes impaired in AD. Some studies have suggested that information in semantic memory becomes eroded, such that category knowledge of concepts might be retained but knowledge of semantic attributes might be disrupted (Hodges et al., 1992; Martin & Fedio, 1983). Others have suggested disorganization in attribute knowledge (Grober et al., 1985) or deterioration in the associative structure between concepts (Abeysingh et al., 1990). Other potential sources of difficulty that have been suggested are impaired access to semantic knowledge (Nebes et al., 1984) and impaired access to lexical knowledge (Funnell & Hodges, 1991).

GENERAL CONCLUSIONS

We have seen in our review of studies of naming in normal aging and AD that the naming problems of AD are in many respects different from those of the naming decline seen in normal aging. Patients with AD generally have much more severe naming problems than normal elderly people do. They are less able to utilize cues, and make many more off-target errors than normal elderly. However, the question remains whether these differences might be just a matter of degree. Studies investigating consistency of naming performance across time or across task have not convincingly indicated that patients with AD show a loss of information from semantic memory. Furthermore several studies using on-line processing techniques (Nebes, 1989; Nebes & Brady, 1988, 1990; Nebes et al., 1984), suggest that aspects of semantic memory organization may remain intact in patients with AD despite poor naming performance.

We have also seen that the overt expression of the naming problem in patients with AD is dependent on dementia severity. It is also likely that individual differences in brain organization premorbidly as well as individual differences in patterns of brain pathology related to AD result in variable presentations of naming impairments across subjects. In recent years, investigators have begun to recognize different subtypes of AD (e.g., Liebson & Albert, 1994; Mayeux, Stern, & Spanton, 1985). For example, some patients may display significant language problems as early signs of their dementia; others may show primary difficulties in visuospatial functioning. In group studies of patients with AD, these individual differences are often eliminated, possibly masking important findings with respect to naming impairment and its underlying mechanisms in these patients.

In contrast, the underlying cause of the naming impairment in normal aging is less in dispute. Many researchers have concluded that the mild anomia associated with normal aging results from a lexical access problem, while information in semantic memory remains intact.

ACKNOWLEDGMENTS

This work was supported by a grant from the Medical Research Service of the Department of Veterans Affairs. We wish to thank the many research assistants in the Language in the Aging Brain laboratory who have helped us over the years. We also gratefully acknowledge the contribution of our research subjects.

REFERENCES

Abeysingh, S. C., Bayles, K. A., & Trosset, M. W. (1990). Semantic memory deterioration in Alzheimer's subjects: Evidence from word association, definition, and associate ranking tasks. *Journal of Speech and Hearing Research, 33,* 574–582.

Albert, M. S., Heller, H. S., & Milberg, W. (1988). Changes in naming ability with age. *Psychology and Aging, 3*(2), 173–178.

Au, R., Joung, P., Nicholas, M., Kass, R., Obler, L. K., & Albert, M. L. (1995). Naming ability across the lifespan. *Aging and Cognition, 2*(4), 300–311.

Barth, C., Nicholas, M., Au, R., Obler, L. K., & Albert, M. L. (1996). *Verb naming in normal aging*. Unpublished manuscript.

Bayles, K. A., & Kaszniak, A. W. (1987). *Communication and cognition in normal aging and dementia*. Austin, TX: Pro-Ed.

Bayles, K., & Tomoeda, C. (1983). Confrontation naming impairment in dementia. *Brain and Language, 19*, 98–114.

Bayles, K. A., Tomoeda, C. K., Kaszniak, A. W., & Trosset, M. W. (1991). Alzheimer's disease effects on semantic memory: Loss of structure or impaired processing? *Journal of Cognitive Neuroscience, 3*(2), 166–182.

Bayles, K. A., Tomoeda, C. K., & Trosset, M. W. (1990). Naming and categorical knowledge in Alzheimer's disease: The process of semantic memory deterioration. *Brain and Language, 39*, 498–510.

Borod, J. C., Goodglass, H., & Kaplan, E. (1980). Normative data on the Boston Diagnostic Aphasia Examination, Parietal Lobe Battery, and the Boston Naming Test. *Journal of Clinical Neuropsychology, 2*, 209–215.

Bowles, N., Obler, L. K., & Albert, M. L. (1987). Naming errors in healthy aging and dementia of the Alzheimer type. *Cortex, 23*, 519–524.

Bowles, N. L., Obler, L. K., & Poon, L. W. (1989). Aging and word retrieval: Naturalistic, clinical, and laboratory data. In L. W. Poon, D. C. Rubin, & B. A. Wilson (Eds.), *Everyday cognition in adulthood and late life* (pp. 244–264). Cambridge, UK: Cambridge University Press.

Bowles, N. L., & Poon, L. W. (1985). Aging and retrieval of words in semantic memory. *Journal of Gerontology, 40*(1), 71–77.

Brown, A. S., & Mitchell, D. B. (1991). Age differences in retrieval consistency and response dominance. *Journal of Gerontology, 46*(6), 332–339.

Burke, D. M., MacKay, D. G., Worthley, J. S., & Wade, E. (1991). On the tip of the tongue: What causes word finding failures in young and older adults? *Journal of Memory and Language, 30*, 542–579.

Caramazza, A., & Hillis, A. E. (1991). Lexical organization of nouns and verbs in the brain. *Nature (London), 349*, 788–790.

Chertkow, H., & Bub, D. (1990). Semantic memory loss in Alzheimer-type dementia. In M. F. Schwartz (Ed.), *Modular deficits in Alzheimer-type dementia* (pp. 207–244). Cambridge, MA: M.I.T. Press.

Chertkow, H., Bub, D., & Seidenberg, M. (1989). Priming and semantic memory loss in Alzheimer's disease. *Brain and Language, 36*, 420–446.

Cohen, G., & Faulkner, D. (1986). Memory for proper names: Age differences in retrieval. *British Journal of Developmental Psychology, 4*, 187–197.

Cooper, P. V. (1990). Discourse production and normal aging: Performance on oral picture description tasks. *Journal of Gerontology, 45*(5), 210–214.

Cormier, P., Margison, J. A., & Fisk, J. D. (1991). Contribution of perceptual and lexical-semantic errors to the naming impairment in Alzheimer's disease. *Perceptual and Motor Skills, 73*(1), 175–183.

Cronin-Golomb, A., Corkin, S., & Rizzo, J. F. (1991). Visual dysfunction in Alzheimer's disease: Relation to normal aging. *Annals of Neurology, 29*(1), 41–52.

Damasio, A. R., & Damasio, H. (1992). Brain and language. *Scientific American, 267*, 88–95.

Damasio, A. R., & Tranel, D. (1993). Nouns and verbs are retrieved with differently distributed neural systems. *Proceedings of the National Academy of Sciences, USA, 90*, 4957–4960.

Francis, W. N., & Kucera, H. (1982). *Frequency analysis of English usage: Lexicon and grammar*. Boston: Houghton-Mifflin.

Funnell, E., & Hodges, J. R. (1991). Progressive loss of access to spoken word forms in a case of Alzheimer's disease. *Proceedings of the Royal Society of London—Biology, 243*(1307), 173–179.

Gentner, D. (1981). Verb semantic structures in memory for sentences: Evidence for componential representation. *Cognitive Psychology, 13,* 56–83.

Glaser, W. R. (1992). Picture naming. *Cognition, 42,* 61–105.

Goldstein, F. C., Green, J., Presley, R., & Green, R. C. (1992). Dysnomia in Alzheimer's disease: An evaluation of neurobehavioral subtypes. *Brain and Language, 43,* 308–322.

Goodglass, H., & Kaplan, E. (1983). *The Boston Diagnostic Aphasia Examination.* Philadelphia: Lea & Febiger.

Goulet, P., Ska, B., & Kahn, H. J. (1994). Is there a decline in picture naming with advancing age? *Journal of Speech and Hearing Research, 37,* 629–644.

Grober, E., Buschke, H., Kawas, C., & Fuld, P. (1985). Impaired ranking of semantic attributes in dementia. *Brain and Language, 26,* 276–286.

Heller, R. B., & Dobbs, A. R. (1993). Age differences in word finding in discourse and nondiscourse situations. *Psychology and Aging, 8*(3), 443–450.

Henderson, V., Mack, W., Freed, D. M., Kempler, D., & Andersen, E. S. (1990). Naming consistency in Alzheimer's disease. *Brain and Language, 39,* 530–538.

Hodges, J. R., Salmon, D. P., & Butters, N. (1992). Semantic memory impairment in Alzheimer's disease: Failure of access or degraded knowledge? *Neuropsychologia, 30*(4), 301–314.

Huff, F. J., Corkin, S., & Growdon, J. H. (1986). Semantic impairment and anomia in Alzheimer's disease. *Brain and Language, 28,* 235–249.

Huff, F. J., Mack, L., Mahlmann, J., & Greenberg, S. (1988). A comparison of lexical-semantic impairments in left hemisphere stroke and Alzheimer's disease. *Brain and Language, 34*(2), 262–278.

Humphreys, G. W., Riddoch, M. J., & Quinlan, P. T. (1988). Cascade processes in picture identification. Special issue: The cognitive neuropsychology of visual and semantic processing of concepts, *Cognitive Neuropsychology, 5*(1), 67–104.

Kaplan, E., Goodglass, H., & Weintraub, S. (1976). *Boston Naming Test: Experimental Edition.* Boston: VA Medical Center.

Kaplan, E., Goodglass, H., & Weintraub, S. (1983). *Boston Naming Test.* Philadelphia: Lea & Febiger.

Kemper, S. (1991). Language and aging: What is "normal aging"? *Experimental Aging Research, 17*(2), 99.

Kirshner, H. S., Webb, W. G., & Kelly, M. P. (1984). The naming disorder of dementia. *Neuropsychologia, 22,* 23–30.

Klatsky, R. L. (1988). Theories of information processing and theories of aging. In L. L. Light & D. B. Burke (Eds.), *Language, memory, and aging* (pp. 1–16). Cambridge, UK: Cambridge University Press.

Kohn, S. E., Lorch, M. P., & Pearson, D. M. (1989). Verb finding in aphasia. *Cortex, 25,* 57–69.

LaBarge, E., Balota, D. A., Storandt, M., & Smith, D. S. (1992). An analysis of confrontation naming errors in senile dementia of the Alzheimer type. *Neuropsychology, 6*(1), 77–95.

LaBarge, E., Edwards, D., & Knesevich, J. W. (1986). Performance of normal elderly on the Boston Naming Test. *Brain and Language, 27,* 380–384.

Le Dorze, G., & Durocher, J. (1992). The effects of age, educational level, and stimulus length on naming in normal subjects. *Journal of Speech Language Pathology and Audiology, 16*(1), 21–29.

Liebson, E., & Albert, M. L. (1994). Cognitive changes in dementia of the Alzheimer's type. In D. B. Calne (Ed.), *Neurodegenerative diseases* (pp. 615–629). Philadelphia: W. B. Saunders.

Light, L. L., & Burke, D. B. (1988). Patterns of language and memory in old age. In L. L. Light & D. B. Burke (Eds.), *Language, memory, and aging* (pp. 244–272). Cambridge, UK: Cambridge University Press.

Madden, D. J., Pierce, T. W., & Allen, P. A. (1993). Age-related slowing and the time course of semantic priming in visual word identification. *Psychology and Aging, 8*(4), 490–507.

Martin, A., & Fedio, P. (1983). Word production and comprehension in Alzheimer's disease: A breakdown of semantic knowledge. *Brain and Language, 19,* 124–141.

Mayeux, R., Stern, Y., & Spanton, S. (1985). Heterogeneity in dementia of the Alzheimer type: evidence of subgroups. *Neurology, 35,* 453–461.

Maylor, E. A. (1990). Recognizing and naming faces: Aging, memory retrieval, and the tip of the tongue state. *Journal of Gerontology, 45*(6), 215–226.

Maylor, E. A., & Valentine, T. (1992). Linear and nonlinear effects of aging on categorizing and naming faces. *Psychology and Aging, 7*(2), 317–323.

Miceli, G., Silveri, M. C., Nocentini, U., & Caramazza, A. (1988). Patterns of dissociation in comprehension and production of nouns and verbs. *Aphasiology, 2,* 351–358.

Miceli, G., Silveri, M. C., Villa, G., & Caramazza, A. (1984). On the basis for the agrammatic's difficulty in producing main verbs. *Cortex, 20,* 207–220.

Myerson, J., Ferraro, F. R., Hale, S., & Lima, S. D. (1992). General slowing in semantic priming and word recognition. *Psychology and Aging, 7*(2), 257–270.

Nebes, R. (1989). Semantic memory in Alzheimer's disease. *Psychological Bulletin, 106,* 377–394.

Nebes, R. D., & Brady, C. B. (1988). Integrity of semantic fields in Alzheimer's disease. *Cortex, 24,* 291–299.

Nebes, R. D., & Brady, C. B. (1990). Preserved organization of semantic attributes in Alzheimer's disease. *Psychology and Aging, 5*(4), 574–579.

Nebes, R. B., Martin, D. C., & Horn, L. C. (1984). Sparing of semantic memory in Alzheimer's disease. *Journal of Abnormal Psychology, 93*(3), 321–330.

Neils, J., Brennan, M. M., Cole, M., Boller, F., & Gerdeman, B. (1988). The use of phonemic cueing with Alzheimer's disease patients. *Neuropsychologia, 26*(2), 351–354.

Nicholas, L. E., Brookshire, R. H., MacLennan, D. L., Schumacher, J. G., & Porrazzo, S. A. (1989). The Boston Naming Test: Revised administration and scoring procedures and normative information for non-brain-damaged adults. In T. E. Prescott (Ed.), *Clinical aphasiology* (vol. 18, pp. 103–115). Austin, TX: Pro-ed.

Nicholas, M., Obler, L. K., Albert, M. L., & Goodglass, H. (1985). Lexical retrieval in healthy aging. *Cortex, 21,* 595–606.

Nicholas, M., Obler, L. K., Au, R., & Albert, M. L. (1996). On the nature of naming errors in aging and dementia: A study of semantic relatedness. *Brain and Language, 54,* 184–195.

Obler, L. K. (1980). Narrative discourse style in the elderly. In L. K. Obler & M. L. Albert (Eds.), *Language and communication in the elderly* pp. 75–90. Lexington, MA: D. C. Heath.

Obler, L. K., & Albert, M. L. (1979). *The Action Naming Test (Experimental Edition).* Boston: VA Medical Center.

Obler, L. K., & Albert, M. L. (1984). Language in aging. In M. L. Albert (Ed.), *Clinical neurology of aging*(pp. 245–253). New York: Oxford University Press.

Puglisi, J. T., Park, D. C., & Smith, A. D. (1987). Picture associations among old and young adults. *Experimental Aging Research, 13*(2), 115–116.

Salthouse, T. A. (1988). Effects of aging on verbal abilities: Examination of the psychometric literature. In L. L. Light & D. B. Burke (Eds.), *Language, memory, and aging* (pp. 17–35). Cambridge, UK: Cambridge University Press.

Sandson, J., Obler, L. K., & Albert, M. L. (1987). Language changes in healthy aging and dementia. In S. Rosenberg (Ed.), *Advances in applied psycholinguistics* (Vol. 1, pp. 264–292). New York: Cambridge University Press.

Schwartz, M., Marin, O. S. M., & Saffran, E. M. (1979). Dissociation of language functions in dementia: A case study. *Brain and Language, 7,* 277–306.

Shuttleworth, E. C., & Huber, S. J. (1988). The naming disorder of dementia of Alzheimer type. *Brain and Language, 34,* 222–234.

Smith, S., Murdoch, B., & Chenery, H. (1989). Semantic abilities in dementia of the Alzheimer type. *Brain and Language, 36,* 314–324.

Stern, C., Prather, P., Swinney, D., & Zurif, E. (1991). The time course of automatic lexical access and aging. *Brain and Language, 40,* 359–372.

Thomas, J., Fozard, J., & Waugh, N. C. (1977). Age-related differences in naming latency. *American Journal of Psychology, 90,* 499–509.

Tulving, E. (1983). *Elements of episodic memory.* New York: Oxford University Press.

Van Gorp, W., Satz, P., Kiersch, M. E., & Henry, R. (1986). Normative data on the Boston Naming Test for a group of normal older adults. *Journal of Clinical and Experimental Neuropsychology, 8,* 702–705.

Warrington, E. K. (1975). The selective impairment of semantic memory. *Quarterly Journal of Experimental Psychology, 27,* 635–657.

Wingfield, A., Goodglass, H., & Smith, K. (1990). Effects of word-onset cuing on picture naming in aphasia: A reconsideration. *Brain and Language, 39,* 373–390.

Chapter 8

Treatment of Aphasic Naming Problems

Nancy Helm-Estabrooks

As preceding chapters in this volume make clear, naming problems are a core symptom of aphasia, and all persons with aphasia have naming problems. Of course, the underlying mechanisms responsible for these problems may differ among aphasia syndromes and even within an aphasic individual, but their prevalence and effect on communication make them a natural target for treatment. Furthermore, all of us have experienced word retrieval difficulty and can appreciate how annoying it is to be unable to access a specific word at a specific moment within a communicative exchange. It is perhaps not surprising, therefore, that clinicians want to help even mildly impaired aphasic patients improve their ability to "name". Moreover, those of us who have successfully traveled in a foreign country using a 20- or 30-word vocabulary important to our needs appreciate that by giving our severely impaired patients even a small lexicon of 'key' words we will greatly improve their quality of life. This has led many clinicians to engage in what Holland (1989) refers to as "itsa" therapy. In "itsa" therapy the clinician identifies objects that the patient cannot name to confrontation. She then asks the patient to repeat each name and then to produce the target word again to the prompt "It's a . . . " when shown a pictured or real object. A more elaborate version of the "itsa" approach is to give patients cues about the target word rather than just providing it for them. These cues can take many forms including initial sounds (phonemic cuing) (e.g., "It's a /ka . . . /"), highly predictable sentences frames (sentence completion cuing)— e.g., "You drive a . . . ", or similar sounding words (rhyme cues)— e.g., "It sounds like 'bar.' It's a . . . " We know from several studies (e.g., Pondraza & Darley, 1977; Pease & Goodglass, 1978) that such cuing techniques are quite successful in eliciting target words from aphasic patients. In fact, most aphasiologists know that with the right cues it is possible to elicit correct verbal labels from most aphasic patients. At the same time, however, we may observe the transitory nature of cuing effects. Even after the lapse of only a few minutes or after the immediate presentation of a new item, a patient may be unable to retrieve a name that just moments before was successfully elicited through cuing. This clinical observation was confirmed in a formal study by Patterson, Purell, and Morton (1983), who compared the effects of multiple repetitions and phonemic cuing on object naming. Patients received three sessions of therapy. Although both types of cues had immediate effects, even the stronger effects of phonemic cuing disappeared after only 30 minutes. Patterson and her colleagues concluded that "while confirming that phonemic cuing is effective at the time of its administration, no evidence has been obtained for its longer-term efficacy" (p. 85).

Thus, cued naming appears to have little therapeutic potential. One possible exception to this observation is the technique of "self-cuing," that is, having patients create their own methods for recalling names. Marshall, Neuburger, and Phillips (1992) compared the effects of six sessions of repetition and sentence completion cuing with six sessions of self-generated cuing on aphasic naming of visually novel stimuli. One week after therapy ceased, naming of items trained with repetition and sentence-completion cues had dropped close to baseline. In contrast, performance of items treated with self-cuing, although not as effective during the training tasks, showed less decline after 1 week.

Cuing studies cited thus far have limited training to a few trials that seem to have had little long-term effect on naming ability; but what about extended drills? If a correct label is elicited over and over, will it begin to leave a trace? In time, is the word accessed with greater and greater facility until it is part of the patient's easily available lexicon? To begin to answer these questions we can look back to a 1924 publication by the American psychologist Shephard Ivory Franz. Franz (1924) was the first person to systematically explore the effects of extended drills on aphasic naming performance. His "Studies in Re-Education" of three long-term aphasic patients are notable for their detailed documentation of his procedures and the patients' responses. During his experiments Franz discovered some basic truths about naming drills. He learned, for example, that patients may fluctuate in their performance from trial to trial, naming an item one time and not the next; that patients who seem to have the same kind of naming problems may show different patterns of response; and that perseveration (the inappropriate repetition of a previous response) is a common byproduct of drills. Franz's purpose in carrying out these studies was to "demonstrate the course of relearning" (p. 350). He did not theorize about the underlying nature of his patients' naming problems; he simply described and quantified them. He did not discuss the possible mechanisms responsible for improvement; he just documented his patients' responses to the naming drills.

Franz was followed by others in taking a rather perfunctory approach to the treatment of aphasic naming problems. In 1951 Wepman advised that therapy for any speech output problem in aphasia begin with treatment of naming problem through the elicitation of a small number of words with various multimodality cues. One patient underwent 3 months of intense drills before naming any item without cues. Similarly, Stewart (1966) used multiple cues to drill her patient for approximately 75 hr after which she could "recite by rote the 108 nouns contained in the nucleus vocabulary" (p. 770). Unfortunately, she appeared unable to use these words in communicative situations. Such cases suggest that the payoff from cued-naming drills is poor. Researchers of today, however, might argue that these early studies were not well controlled either in the application of treatment techniques or the measurements of treatment effects.

In 1973 a landmark study of naming therapy was conducted by Weigl-Crump and Koenigsknecht. One notable feature of their study was the investigators' effort to answer a theoretical question centered on the long-standing issue of whether aphasia represents a competency or a performance deficit. In treating naming problems, Weigl-Crump and Koenigsknecht were trying to determine if the anomic component of aphasia represents a loss of efficiency in retrieving words from the lexical store, or a reduction in that store itself. They hypothesized that generalization of improved naming skills for trained items to untrained items would be evidence of a retrieval deficit. Another notable feature of this study was their effort to control extrinsic treatment variables, such as time postonset, and instrinsic variables, such as the frequency of occurrence and semantic categories of target words. A multimodal stimulation approach of the type recommended by Schuell, Jenkins, and Jimenez-Pabon (1969) was used to elicit responses from four patients with moderate naming problems. Increases in the average number of drilled items named without cues after therapy was interpreted as indicating significant improvement. More importantly, however, Weigl-Crump and Koenigsknecht also found improvement in naming nondrilled items. They stated that their experimental findings support the argument that anomia represents a retrieval problem rather than a reduction in patients' lexical store. Unfortunately, subsequent studies (e.g., Davis & Pring, 1991; Marshall, Pounds, White-Thompson, & Pring, 1990) have failed to find generalized effects to nondrilled items. Thus, the issue of retrieval versus loss of language in aphasia has not been settled through studies of naming therapy.

A review of published approaches to treating aphasic naming problems indicates that the choice of therapy method is greatly influenced by the profession and training of the person doing the research. For example, Stewart (1966), whose work was cited above, was a speech therapist working in a school system where drills often are used successfully with developmental articulation problems. So, when asked to treat an aphasic adult, she may have turned to simple naming drills without further consideration. In contrast, clinicians whose experience emphasized the neurological foundations for language have based their treatments on theories of functional brain mechanisms. Examples of these more neurological methods are the functional reorganization approach of Luria (1970) and the closely allied deblocking approach of Weigl (1968). Clinical researchers with training in neurolinguistic and cognitive psychology tend to base their aphasia treatment approaches on cognitive models of normal language, for example, Howard, Patterson, Franklin, Orchard-Lisle, and Morton (1985) described their naming therapy in terms of the "logogen" model of semantic selection proposed by Morton (1969). In fact, most current approaches to treatment of aphasic naming disorders can be grouped under these two headings: functional reorganization and deblocking methods, and cognitive-model-driven approaches. We will use these headings, therefore, to organize the following review of naming therapies.

FUNCTIONAL REORGANIZATION AND DEBLOCKING THERAPIES FOR NAMING

Intersystemic and Intrasystemic Reorganization

In a 1947 monograph Alexander Luria described his approaches to aphasia rehabilitation based on his experiences with brain-injured Russian soldiers in the Second World War. In subsequent years this text underwent several revisions and the English version of "Traumatic Aphasia" (Luria, 1970), is still available. In it Luria presented general principles and specific approaches for restoring disturbed brain functions through the reorganization of cortical processes. Luria believed that "in man almost any cortical area can acquire new functional significance and thus may be incorporated into almost any functional system" (p. 382). As an example of the application of this principle, Luria described a patient with an occipital lobe lesion who was unable to recognize letters, but, by tracing the contours of letters with movements of his finger, was able to "synthesize" and recognize letters. Thus, the patient developed a new functional system that allowed him to read. This compensatory approach to reading is an example of *intersystemic reorganization,* because a different functional system (finger movements) was used to compensate for a visual-based reading problem.

A recent study by Hanlon, Brown, and Gerstmann (1990) can be regarded as an example of an intersystemic approach to facilitating naming skills. These investigators sought to determine the effects of unilateral gestural

movements (pointing and making a fist) on the confrontation-naming performance of severely aphasic patients with either Broca's, Wernicke's,
global, transcortical motor or anomic aphasia. They predicted that performance of nonfluent patients would improve when they employed the
proximal (shoulder) musculature of the right arm while trying to name pictures. Pictures were placed in patients' right peripheral visual field (at
about a 60-degree angle), so that patients had to rotate their head and/or
trunk to see the picture. This rotation usually elicited an asymmetric tonic
neck reflex that temporarily overcame the patients' right hemiplegia so
that they could point to the picture while naming. In the second condition,
pictures were placed in front of the patients, and they were told to make a
fist with their right hand while naming. In the third condition, pictures
were placed in patients' left peripheral visual field, and left-hand pointing
accompanied their naming attempts. Hanlon and his colleagues found that
naming was significantly better during the right-pointing condition for
nonfluent patients which they attributed to a "microgenic process model"
(p. 311). According to this model, the use of the hemiplegic right arm for
pointing activated the archaic proximal motor system with resultant access
to anterior action systems and stimulation of vocal articulation. This approach to stimulating naming through the activation of a different motor
system may have therapeutic merit, but a controlled treatment study must
be carried out to determine whether long-term improvements in naming
skills would result from this form of *intersystemic reorganization.*

Luria also discussed *intrasystemic reorganization,* which involves the use
of different elements or levels of activity within the disturbed function. One
example of intrasystemic reorganization is the elicitation of reactive or involuntary speech from the patient, which, then, is brought to a conscious or
voluntary level, and finally rehearsed to improve its use. This approach to
restoring speech skills was employed by Vignolo (1964) in his study of the
effects of aphasia therapy. In describing his techniques he stated, "First an
automatic way to elicit a correct response is found, and the response is then
tentatively elicited in more and more voluntary ways" (p. 349).

Along similar lines, Helm and Barresi (1980) studied an intrasystemic
approach to improving naming skills with a method we called Voluntary
Control of Involuntary Utterances (VCIU). The subjects were three severely aphasic patients whose attempts to communicate were mostly limited
to a few real-word stereotypic expressions with little to no ability to produce correct names to confrontation. At 2-3 months postonset they had
shown poor response to other forms of therapy. A review of one case (Mr.
N.) will illustrate how the VCIU method was carried out.

VCIU with Mr. N. began with identification of all substantive words he
had uttered during a comprehensive language exam regardless of their
correctness. For example, on a repetition subtest he said "kiss" for *hammock*

and "two" for *W*. On a confrontation-naming subtest he said "die" for *chair*. Although he scored no points on the oral reading subtest, when presented with his own words (kiss, two, die) he was able to read them aloud. Next, he was presented with two new emotion-laden words (*love* and *war*), which he also read correctly. Then, the printed word *father* elicited "mother," so it was replaced with *mother* which he read as "mama." Mother was replaced with *mama*, which was read correctly. Proceeding in this manner, eight words that Mr. N. could reliably read aloud were identified during the first session. Each word was written on a separate index card and pictures representing these words were drawn on the reverse sides. After reading the words aloud, Mr. N. was asked to turn over the cards and "name" the pictures. The next step was for him to name the pictures without first reading the target words. In this manner, his vocabulary was expanded. By the fifth session, his wife reported some new words he produced at home and these were incorporated into his VCIU training.

At the end of 3 months of VCIU therapy, Mr. N. had a functional vocabulary of 100 words, such as "eggs," "juice," "coffee." The other two patients studied showed similar levels of response, and all three made significant gains on posttherapy tests of confrontation naming.

As compelling as the VCIU results might appear, it should be pointed out that the 1980 study was not well controlled by today's research standards. Although the three patients had not responded to other therapies administered during the first 2–3 months postonset, some of the gains may have resulted from delayed natural recovery. Baseline probes over a 1-month period might have helped determine whether "spontaneous" improvement was still occurring. It was not possible to delay therapy, however, because these patients had been admitted to the aphasia unit specifically for treatment. One solution to this "immediate treatment" problem might have been to use an ABAB treatment design in which VCIU was systematically alternated with another therapy approach. Despite the methodological flaws of the study, VCIU seems to have merit in view of each patient's quick response to this approach after months of showing no improvement in naming with other approaches. Their quick response to VCIU also suggests that lexical items were not being taught, but rather old, established vocabulary items were being "deblocked." A full description of the VCIU method is found in Helm-Estabrooks and Albert (1991).

Deblocking Methods

A core assumption of "deblocking" is that the capacity for language performance is not entirely lost in most aphasic patients; that a "blocked" function can be accessed through a different, more intact channel. Thus, performance within a more preserved function is elicited just before probing with the disturbed function. In the case of VCIU the more preserved function was oral reading of words that were initially uttered as verbal errors on other tasks. The disturbed function was oral naming. So, oral reading of these words was used to deblock their purposeful use in a confrontation-naming task. If other steps were added between oral reading and naming we would have an example of what Weigl referred to as "chain deblocking." His own example of chain deblocking was that of a patient who could not name a picture of a *boy*, repeat this word, or write it from dictation. But, he could read the word "boy" aloud, and he understood what it meant. The "chain" of deblocking in this case proceeded from oral reading, to repetition, to writing the dictated word, to naming the picture.

It is important to note that although Weigl describes several cases in which deblocking techniques were used with apparent success, he reported no formal studies of his methods. The use of Osman-Sagi's (1993) broader definition of deblocking as a way "to liberate the functions which are inhibited but not lost" (p. 85), however, allows us to discuss what could be regarded as a formal "deblocking" study (Helm-Estabrooks, Emery, & Albert, 1987). The "blocker" in this study was recurrent perseveration (the inappropriate repetition of all or part of a previous word), a phenomenon that occurs commonly in aphasic naming attempts. The "blocked" skill was confrontation naming. Three patients were identified as having moderate to severe perseveration and low scores on the confrontation naming subtest of the Boston Diagnostic Aphasia Examination (BDAE) (Goodglass & Kaplan, 1983). The method for controlling perseveration was the following: (1) bring perseverative naming responses to the patients' level of awareness; (2) help them inhibit those perseverative responses; and (3) train them to ask for cues for accessing and producing correct naming responses. A more complete description of this approach, called TAP (Treatment of Aphasic Perseveration), is found in Helm-Estabrooks and Albert (1991). All three patients were treated with an ABAB design that alternated TAP with some other form of treatment in five session blocks. The dependent variable was the BDAE confrontation-naming subtest, which was readministered after each training block. Training items represented the same semantic categories as the BDAE (i.e., objects, letters, geometric forms, actions, numbers, colors, and body parts). For each patient, the TAP program was significantly more effective than the alternative approach in reducing perseveration, and as perseveration decreased, naming scores on the BDAE subtest increased.

Thus, functional reorganization and deblocking methods for restoring naming skills are based on the assumption that the ability to name is not "lost," but rather that the primary channels for naming are disrupted.

Through identification and use of alternative channels, or through elimi-
nation of factors that are "blocking" primary channels, the ability to name
is, to some extent, restored. These methods are not based on a cognitive
neuropsychological model of naming, but on an understanding of spared
and impaired brain pathways and functions and on ideas about the nature
of the rehabilitation process. In the 1980s, however, cognitive psychologists
became interested in aphasic naming problems, and the result was a flur-
ry of published studies of cognitive theory-driven treatment approaches,
many of which were authored by British investigators.

COGNITIVE THEORY-DRIVEN APPROACHES
TO TREATMENT OF NAMING

Cognitive neuropsychologists became interested in the treatment of
aphasic language as a means of advancing their understanding of the cognitive processes responsible for normal language understanding and production. Using language-processing models they identified the levels at
which aphasic subjects' performance broke down, then aimed treatments
at the processes presumed to be impaired. It is important to note that unlike reorganization and deblocking approaches, which are based on theories of therapy or "how to fix deficits" (Holland, 1994, p. 276), cognitive
neuropsychological approaches are based on theories about the nature of
these deficits vis-à-vis normal language processes. In this respect studies
of cognitive model-driven therapies seem to have broken new ground in
the area of aphasia rehabilitation. But, modern cognitive psychologists are
not the first to offer theories about the underlying nature of specific aphasic deficits. Goldstein (1942), for example, thought that severe anomia
("amnesic aphasia") resulted from "impairment of the abstract attitude."
He believed that naming depends on one's ability to understand that
"when speaking 'table' we do not mean a special, given table with all the
accidental properties, but we mean 'table' in general" (p. 155). He con-
tended that patients with anomia had lost their abstract attitude and were
dealing with each item individually, in a concrete manner. Based on his
theory of anomia, and on failed attempts to treat patients with random
naming drills, Goldstein thought a better approach might be to reteach
names of objects in categories with a "natural relationship" (e.g., clothing).
Thus, in modern neuropsychological terms, Goldstein was suggesting that
some aphasic patients have problems within their semantic system and
that the problem might be overcome by reinforcing semantic links between
items. This approach has been used in a number of recent studies of cognitive model-driven therapies, a few of which are described below.

One of the clearer explanations of the use of a neurocognitive approach to treating aphasic naming problems is that of Nettleson and Lesser (1991). It will be described here in some detail to illustrate how cognitive models may be applied to treatment. Nettleson and Lesser's model for naming described a semantic system that feeds into a phonological output lexicon which, in turn, feeds into a phonological assembly buffer with the final stage involving "the more peripheral operations of phonetic planning and neuromuscular execution" (p. 143). Nettleson and Lesser presented criteria for determining the level of impairment within the model's naming process for each of their six aphasic patients. They ruled out peripheral level problems by excluding patients with dysarthria, and identified naming problems according to abnormal performance on the Boston Naming Test (BNT) (Kaplan, Goodglass, & Weintraub, 1983). Their criteria for determining that a naming disturbance resulted from *impairment of the semantic system* included the following:

1. Below normal performance on semantic decision tasks of the Psycholinguistic Assessment of Language Processing in Aphasia (PALPA) (Kay, Lesser, & Coltheart, 1990).
2. Most BNT errors were semantic paraphasias.
3. Phonemic cues for semantic associates of target words usually elicited these words and patients accepted them as correct.

The criteria for determining that a naming disturbance arose from *phonological output lexicon problems* were as follows:

1. No difficulty with auditory word–picture matching on the PALPA.
2. The majority of BNT errors were anomic circumlocutions.
3. Percentile scores on BDAE subtests of repetition were at or above the percentile scores for auditory comprehension.

The three criteria used to determine that the naming disorder related to *disturbance within the phonological output buffer* were as follows:

1. Normal scores on the PALPA auditory-word picture-matching task.
2. The majority of BNT naming errors were phonemic paraphasias.
3. The percentile scores earned on BDAE repetition subtests were below those of auditory comprehension subtests.

Using these criteria, they identified two patients with each category of naming impairment. The naming skills of these patients were then assessed with a set of 300 line drawings, and 100 of each patient's unnamed items were targeted. Fifty of these unnamed items were randomly allocated to a no-treatment set and 50 to a treatment set.

Before therapy began, Nettleson and Lesser obtained eight measures of naming at weekly intervals using untreated items in random groups of 25 items each. In the ninth week each patient's ability to name all 300 items was again assessed, and treatment began at the rate of two sessions a week for 8 weeks. Four patients received "model-appropriate" treatment (i.e., semantic therapy for the two patients with semantic system problems and phonological therapy for the two with phonological lexicon problems). The two patients with phonological assembly problems were given "model-inappropriate" therapy (i.e., semantic therapy). Semantic therapy involved written and spoken word-picture matching, yes–no judgments about categories and attributes, and sorting by categories. In none of these tasks were patients asked to name items. Phonological therapy consisted of the repetition of picture names, rhyme judgments, and naming with progressive cues.

After 8 weeks of therapy three of the four patients who received "model-appropriate" treatment showed significant improvement in naming trained items. The two patients with phonological assembly problems did not improve with the "model-inappropriate" semantic therapy. The semantic problem–semantic therapy group showed some generalized improvement to the BNT. Nettleson and Lesser stated, however, that "the design of the study limits the firm conclusions that can be drawn" (1991, p. 153). They pointed out that they did not determine whether the two patients receiving model-inappropriate therapy would have improved with model-appropriate treatment for their phonological assembly buffer problems. Also, they did not test whether the three subjects who improved with model-appropriate treatment would have ceased to improve if, subsequently, they were treated with model-inappropriate therapy. They also pointed out that one of the patients receiving model-appropriate therapy showed no improvement, thereby failing to support the treatment model. Nettleson and Lesser (1991) concluded that "the tendency may be to apply any model in a simplistic manner, and to overlook the possibility that an aphasic patient may have problems at more than one level of the naming process" (p. 155).

Another team of investigators that used a semantic therapy approach based on a cognitive model of naming was that of Marshall, Pounds, White-Thompson, and Pring (1990). The intervention method they used with three single cases involved matching pictures to spoken words and distinguishing correct words from semantic distractors. They assumed that these tasks would reinforce links between the semantic system and the phonological output lexicon. For a group of seven different patients, the method used was the same except that written words were used so that patients could practice at home. Posttherapy measures showed that two of the three single cases significantly improved in their ability to name trained items. The third patient showed improvement on both trained and untrained items, but this finding is difficult for the reader to interpret in view of the obviously erroneous statement that the patient had a stroke in

May of 1987 and began treatment in October of 1986. For the group of seven, significant improvement occurred on both treated items and untreated, semantically related items. Interestingly, Marshall and his colleagues concluded with the statement that "criticisms of efficacy research of this kind miss an important point. A distinction should be made between research that seeks to show that therapy works and research that tries to evaluate the methods that might be used in therapy" (p. 182).

Another study (Davis & Pring, 1991) evaluated methods that might be used for treating naming problems. The first method (semantic) required patients to match target pictures to one of four semantically related words and then to repeat or read the target word aloud. The second method differed only in so far as the three distractors words were semantically unrelated to the targets. The third method (phonological) merely required patients to repeat the names of items spoken in the presence of pictured targets. Davis and Pring hypothesized that if exposure to semantic elements is more important to treatment of naming, the first method would be most helpful, followed by the second method, then the third. They further hypothesized that if the critical therapy component is just repeated exposure to pictures and their names, then no differences would be found among the three approaches. Seven patients who were 5 months to 3.5 years postonset of aphasia received a minimum of 10 therapy sessions over about 5 weeks. It should be noted that three patients were simultaneously receiving group therapy and one was receiving another form of individual treatment. Davis and Pring report that all three treatment approaches resulted in significant improvement in naming treated items, and this improvement lasted for up to 6 months. Furthermore, improvement occurred on unrelated, but not related, distractor items. They pointed out that multiple repetitions of target words were used in their study, and they conjectured that "the form of the task is less important than the number of times it is carried out" (p. 143). Thus, the authors of this particular model-driven naming therapy study bring us full circle in speculating as to the therapeutic effects of repeated drilling.

CONCLUDING THOUGHTS ON NAMING THERAPY

The notion that repeated trials are important to retraining aphasic naming problems was introduced at least 70 years ago, and this idea has not disappeared with the publication of recent, more sophisticated, neurocognitive model-driven therapy studies. As Holland (1994) pointed out, neuropsychologists, as well as speech-language pathologists, seem to believe that language drills can fix naming problems. She stated further that the questionable assumption seems to be that patients must practice what is hard to do (or cannot be done at all) until they get it right. It is my observation that the naming problems of patients with aphasia can be quite elusive and difficult to treat in any direct manner. No matter what form cued drills take they appear to be of such little lasting benefit that they are not worth the expenditure of time, effort, and money. At least one difficulty presented by direct treatments of naming problems is that patients are expected to produce a specific word within a short time frame. The therapies, themselves, not to mention the naming tests used to measure treatment effects, are confrontational in manner. Holland (1989) proposed that a better approach might be to engage patients in "nonconfrontational" tasks that do not require them to produce explicit names for explicit pictures. In keeping with this notion, the neuropsychologist Edith Kaplan (personal communication, April 1987) observed that aphasic patients often produce better language incidently during ostensibly nonverbal, cognitive tasks than when they are required to verbalize. For example, one of her patients, who earned no points on the BNT, began to name objects to himself as he tried to solve a picture arrangement task. When she realized what he was doing, Dr. Kaplan pointed out that he was "naming" the pictures. Unfortunately, by bringing the patient's "incidental" naming to his attention he no longer could name the objects; verbalization had now become an intentional task. Certainly there is a message in this anecdote for all of us who seek to treat aphasic naming disorders. First, it is clear that Kaplan's patient had not "lost" his lexicon but, rather, that he had lost his ability to access this lexicon intentionally. This observation echoes Hughling Jackson's description (1878) of aphasia as a disorder of the ability to use language intentionally. Jackson explained that the higher the propositional value of a task, the less likely an aphasic patient will respond correctly. So, perhaps the course to pursue in treating aphasic naming problems is, as Holland (1994) suggested, one that incorporates a variety of nonconfrontational tasks. For example, there is some evidence to suggest that word retrieval improves when aphasic patients are asked to concentrate on drawing pictures rather than on talking (Lyon & Helm-Estabrooks, 1987). In any case, it seems that little will be lost if we set cued naming drills aside and begin to systematically explore nonconfrontational, nonverbal cognitive approaches to treating aphasic naming problems.

ACKNOWLEDGMENTS

The author wishes to thank Richard Curlee, Audrey Holland, and Pelagie Beeson for their helpful editorial comments. This work was supported, in part, by National Institute of Health/National Institute for Deafness and Communication Disorders (NIH NIDCD) grants #DC01409 and DC00081.

REFERENCES

Davis, A., & Pring, T. (1991). Therapy for word-finding deficits: More on the effects of semantic and phonological approaches to treatment with dysphasic patients. *Neuropsychological Rehabilitation, 1*(2), 135–145.

Franz, S. I. (1924). Studies in re-education. The aphasias. *Journal of Comparative Psychology, 4,* 349–429.

Goldstein, K. (1942). *Aftereffects of brain injuries in war: Their evaluation and treatment.* New York: Grune and Stratton.

Goodglass, H., & Kaplan, E. (1984). *Boston Diagnostic Aphasia Examination.* Philadelphia: Lea and Febiger.

Hanlon, R. E., Brown, J. W., & Gertmann, L. J. (1990). Enhancement of naming in nonfluent aphasia through gesture. *Brain and Language, 38,* 298–314.

Helm, N. A., & Barresi, B. (1980). Voluntary control of involuntary utterances: A treatment approach for severe aphasia. In R. Brookshire (Ed.), *Clinical aphasiology conference proceedings* (pp. 308–315). Minneapolis, MN: BRK.

Helm-Estabrooks, N., & Albert, M. L. (1991). *Manual of aphasia therapy* (pp. 229–238). Austin, TX: Pro-Ed.

Helm-Estabrooks, N., Emery, P., & Albert, M. L. (1987). Treatment of Aphasic Perseveration (TAP) program. *Archives of Neurology, 44,* 1253–1255.

Holland, A. L. (1989, June). *Word retrieval deficits: Some thoughts on assessment and management.* Paper presented at the Academy of Aphasia Meeting, Sante Fe, NM.

Holland, A. L. (1994). Cognitive neuropsychological theory and treatment for aphasia: Exploring the strengths and limitations. *Clinical Aphasiology.* Vol. 22, pp. 275–282.

Howard, D., Patterson, K., Franklin, S., Orchard-Lisle, V., & Morton, J. (1985). The facilitation of picture naming in aphasia. *Cognitive Neuropsychology, 2,* 49–80.

Jackson, J. H. (1878). On affectations of speech from disease of the brain. *Brain, 1,* 304–330.

Kaplan, E., Goodglass, H., & Weintraub, S. (1983). *The Boston Naming Test.* Philadelphia: Lea and Febiger.

Kay, J., Lesser, R., & Coltheart, M. (1990). *Psycholinguistic Assessment of Language Processing in Aphasia* (PALPA). London: Lawrence Erlbaum Associates, Ltd.

Luria, A. R. (1947). *Traumatic aphasia.* Izd. Akad. Ped. Nauk RSFSR. Moscow.

Luria, A. R. (1970). *Traumatic aphasia: Its syndromes, psychology, and treatment.* The Hague: Mouton.

Lyon, J. G., & Helm-Estabrooks, N. (1987). Drawing: Its communicative significance for expressively restricted aphasic adults. *Topics in Language Disorders, 8*(1), 61–71.

Marshall, R. C., Neuburger, S. I., & Phillips, D. S. (1992). Effects of facilitation and cuing on labelling of 'novel' stimuli by aphasic subjects. *Aphasiology, 6*(6), 567–583.

Marshall, J., Pounds, C., White-Thompson, M., & Pring, T. (1990). The use of picture/word matching tasks to assist word retrieval in aphasic patients. *Aphasiology, 4*(2), 167–184.

Morton, J. (1969). The interaction of information in word recognition. *Psychological Review, 76,* 165–178.

Nettleson, J., & Lesser, R. (1991). Therapy for naming difficulties in aphasia: Application of a cognitive neuropsychological model. *Journal of Neurolinguistics 6*(2), 139–157.

Osman-Sagi, J. (1993). Psychological mechanisms of speech rehabilitation in aphasic patients. *Acta Neurochirurgica* (suppl.) *56,* 85–90.

Patterson, K., Purell, C., & Morton, J. (1983). Facilitation of word retrieval in aphasia. In C. Code & D. J. Muller (Eds.), *Aphasia therapy: Studies in language disabilities and remediation* (pp. 76–87). London: Edward Arnold.

Pease, D. M., & Goodglass, H. (1978). The effects of cuing on picture naming in aphasia. *Cortex, 14,* 178–189.

Pondraza, B. L., & Darley, F. L. (1977). Effect of auditory prestimulation on naming in aphasia. *Journal of Speech and Hearing Research, 20,* 669–683.

Schuell, H., Jenkins, J., & Jimenez-Pabon, E. (1969). *Aphasia in adults.* New York: Harper Row.

Stewart, F. J. (1966). A nucleus vocabulary in the therapy of dysphasia: Word finding, naming, and recall—a case report. *Journal American Geriatric Society, 14,* 768–771.

Vignolo, L. A. (1964). Evolution of aphasia and language rehabilitation. A restrospective exploratory study. *Cortex, 1,* 344–367.

Weigl, E. (1968). On the problem of cortical syndromes: Experimental studies. In M. L. Simmel (Ed.), *The reach of mind* (pp. 143–159). New York: Springer.

Weigl-Crump, C., & Koenigsknecht, R. A. (1973). Tapping the lexical store of the adult aphasic: Analysis of the improvement made in word retrieval skills. *Cortex, 9,* 411–418.

Wepman, J. M. (1951). *Recovery from aphasia.* New York: The Ronald Press.

Chapter 9

Summary of the Volume

Harold Goodglass and Arthur Wingfield

In our introductory chapter we described the language disturbances associated with aphasia and the brain areas compromised when these symptoms appear. As we did so, we highlighted four patterns of breakdown in naming. The most common form of anomia is a general degradation affecting both the latency for retrieving names of concepts and restriction of the accessible lexicon, chiefly in terms of the frequency of the words to be produced. A second form of anomia that is observed in some patients is a complete dissociation between normal processing of the input, such as pictures or definitions, and the ability to achieve the phonology or orthographic form of the response. A third form is one in which such dissociations may selectively affect particular input channels. Finally, there is a fourth form in which the naming disorder is selective to a particular category. Well-known examples of such category-specific deficits include human-made artifacts, particular parts of speech, and proper names. In the opening chapter we also reviewed the variety of error types that occur in naming attempts: semantic, phonological, multicomponental, and perseverations. These features represent the problems that are posed for explanations of anomia in anatomic and cognitive terms—explanations that are undertaken by four of the chapters in this volume.

In Chapter 2, Gordon points out that virtually all contemporary models of naming include at least three basic stages: that of visual processing of a stimulus to be named, a stage of semantic appreciation of the object, and a stage of phonological retrieval. Among the models currently competing for acceptance, Levelt's (1989) goes further than any others in proposing substages of semantic processing as well as in phonological activation and implementation. One issue of controversy in attempts to model the naming process concerns how information travels through the processing stream: whether processing must be completed at each stage in turn, whether information spreads ("cascades") from each stage forward as soon as processing begins at that stage, or whether information not only feeds forward but is also fed back upstream in a fully interactive manner. There is also

ANOMIA: Neuroanatomical and Cognitive Correlates

some debate as to whether semantic processing is a necessary intervening stage between visual recognition and phonological retrieval. Current experimental evidence does not exclude the plausibility of any of these models.

In Chapter 1, as a starting point for the later discussion of explanatory models, we reviewed the sequential stage anatomic-functional system described by Geschwind (1969), which followed historically from Wernicke's classical analysis. This model had the appeal of associating major diagnostic aphasia types with lesion sites that were presumed to disable particular steps in the naming process. We also cited experimental evidence, however, that placed in serious question the correspondence between types of naming breakdown and such sequences of anatomic stages.

In Chapter 3 one sees a striking contrast between these classical views of the anatomy of naming, and the evidence presented by Tranel, Damasio, and Damasio. These authors concentrate their efforts on distinguishing object naming along three dimensions: unique versus nonunique entities, entities versus actions, and animals versus tools. In so doing, they associate naming deficits with structures that did not play a part in the earlier anatomic models of naming. For example, they find that naming of unique entities (familiar faces) is selectively associated with injury to the left temporal pole, difficulty in naming animals is associated with injury to the left inferior temporal lobe, and the naming of actions is most frequently impaired with injury to the left premotor area. Larger lesions of the posterior temporal region are likely to affect all categories of name retrieval. The angular gyrus and Wernicke's area, which were important way stations in Geschwind's (1969) stepwise model, do not appear in the data described by Tranel et al. One should not, of course, disregard the significance of the older lesion data in favor of the observations from functional imaging. A full account of the anatamo-physiological processes involved in naming must reconcile the differences arising from different sources of information about brain activity in naming. Part of the answer may lie in the difference between lesions that disrupt pathways versus lesions that directly affect anatomic centers of activity.

By taking their cue from recent observations of category-specific lexical difficulties, Tranel et al. demonstrate that the semantic properties of the target to be named may exert a pull to processing in one portion or another of the language zone. It is important to note in this regard that Tranel et al. emphasize that these areas are not to be regarded as "centers" dedicated to the storage of names for particular object categories. Rather, the implicated structures appear to have a probabilistic predilection for processing name retrieval for concepts of the types cited.

An association between the semantic properties of a target to be named and the anatomy involved is also seen in De Bleser's presentation in Chapter 4. As in Tranel et al.'s chapter, De Bleser's chapter serves as a bridge between the traditional anatomic approaches and more contemporary cognitive analyses of naming disorders. This bridge is made possible through her focus on modality-specific naming disorders. These are disorders that are more closely tied anatomically to input–output systems than are dissociations of semantic categories. Specifically, De Bleser confronts the controversy as to whether semantic knowledge is multiply represented in various sensory systems through which objects may be experienced (e.g., visual, auditory, and tactile representations), or whether object concepts are stored in an abstract, supramodal representation.

Favoring the second possibility, most disorders of object naming appear in the same form regardless of the channel of sensory stimulation. That is, failure to name an object on visual confrontation is paralleled by failure to name it by touch or verbal definition. However, the appearance of purely optic aphasia, acoustic aphasia, or tactile aphasia appears to support the existence of multiple modality-specific semantic stores. De Bleser clarifies the relationship between such modality-specific semantic knowledge and the general lexical–semantic representation of concepts. She points out that in cases of optic aphasia it is common to find errors that are semantically related to the presented object, an observation that is puzzling if the disorder is conceived as simply a disconnection of visual input from the semantic knowledge system.

De Bleser suggests that these phenomena are more understandable in a spreading activation model of naming, in which the structural features of a percept activate a family of visually similar semantic representations. Difficulty in selecting among competing representations may result in accessing the phonology of a concept that is related both visually and semantically to the target. As she indicates, such mutuality between visual and semantic influences is satisfied by a cascade type of spreading-activation model, in which activation at one stage is passed along to the next even while processing continues at the earlier stage. De Bleser points out, however, that there are forms of optic aphasia in which semantic influences are rare. This would indicate that the injury to the system is further upstream, in the transition between early low-level visual analysis and the three-dimensional object concept.

In Chapter 5, Semenza focuses on what Tranel et al. referred to as the naming of "unique entities" (i.e., proper nouns, in contrast to generic object names). Semenza points to occurrences of a double dissociation between the naming of people and the naming of common objects. He notes that the selective inability to retrieve proper nouns is associated with a more general impairment in the ability to retrieve arbitrary associations between word pairs, as in paired association learning of previously unassociated word-to-word, or color-to-digit combinations. On these grounds, Semenza argues for a processing system that is unique to the formation of arbitrary associations, typified by proper name retrieval. In his discussion he treats this dissociation primarily in cognitive terms. He questions the evidence for an anatomical focus for such a system, suggesting, rather, that the system underlying proper name retrieval is widely distributed.

The modality-specific dissociations described by De Bleser can be conceived of as disconnections between a sensory input stage and the semantic stage. However, because of the interaction between sensory system-specific and semantic factors she describes, De Bleser's evidence is preferentially accounted for by a spreading activation approach, rather than by a completely stepwise progression between the stages. The category-specific naming disorders documented by Tranel et al. and by Semenza represent a new challenge for models of name retrieval. Specifically, they call for the examination of the structure of lexical semantics at a level beyond the scope of current models of name retrieval.

Chapters 6 and 7 depart from the search for general models of the cognitive and anatomic basis of the naming process, to focus on naming across the life span. The first of these two chapters addresses problems related to the development of naming in children, and the second chapter deals with the decline of naming proficiency with aging. In addition to illustrating problems of development and deterioration, both extremes of the life span include populations at special risk of disability because of age-linked problems of brain dysfunction. In the case of children, the focus is on evidence for the early differentiation between the hemispheres in the development of language. In the case of adult aging, the focus is on the decline of the cognitive resources upon which the naming process relies.

In Chapter 6 Menyuk traces the changes in the nature of the naming act that parallel cognitive development from infancy to late childhood. Initially the word is linked to a unique referent and only with experience does it come to stand for an entire class of basic objects. Still further maturation is needed before the word can be used in a metaphorical sense. Analyzing the effects of brain injury on naming shows much less in the way of site-related specificity for children than is typically the case in adults. In the case of children, early brain damage involving the left hemisphere results in lower naming scores than right-brain damage, but even right-brain-damaged children are impaired in comparison with their normal counterparts. Brain-injured children are also less likely than adults to show naming disorders that are sharply differentiated from other language deficits; their naming difficulties tend to appear in conjunction with impairments in syntax and written language as well. These findings speak to a functional differentiation that takes time to make its appearance through the early years of development.

Recognizing that one of the early symptoms of Alzheimer's disease is the presence of some degree of anomia, Nicholas, Barth, Obler, Au, and Albert in Chapter 7 approach the issue of naming in pathological states by first examining changes in normal aging. They establish that name retrieval is relatively stable in older adults until the age of 70, at which point there is a loss in retrieval efficiency. This is seen both in reduced access to vocabulary and in longer retrieval latencies. Older adults also show an increased incidence of unresolved tip-of-the-tongue (TOT) states, particularly in attempts to retrieve proper nouns. When probed for partial knowledge of the words, they also appear to have less information than younger adults. On the other hand, elderly adults' word finding is aided by priming with the first sound of the sought-after word, and they also show normal semantic priming.

Nicholas et al. interpret the changes in naming with normal aging in terms of a Transmission Deficit Hypothesis, which attributes these effects to a weakening of connections between semantic and phonological elements in a cognitive network (Burke, MacKay, Worthley, & Wade, 1991). Within this framework, the special vulnerability of proper name retrieval would result from the limited number of semantic connections between proper name representations and other concepts.

Nicholas et al. present a careful review of the evidence on whether the name retrieval difficulty in Alzheimer's disease represents a loss of semantic memory for the concepts represented by particular words. Arguments favoring this view have relied heavily on the consistency with which particular words are failed on repeated testing. Nicholas et al. find that the evidence for such consistency is unconvincing. Although there is no doubt about the degradation of semantic memory in Alzheimer's disease, Nicholas et al. focus on impaired retrieval processes as the primary locus for the naming difficulty in these patients.

In Chapter 8 Helm-Estabrooks considers three approaches to the treatment of naming disorders. The first approach is based primarily on repetitious drills, the second approach relies on residual intact channels to restructure name retrieval, and the third approach is to attempt to diagnose the naming disorder in terms of a model, and then to aim treatment at the processing links that are presumed to be damaged. Drilling techniques usually depend on the use of a facilitating cue, such as providing the first sound, to lead the patient to the desired response. As Helm-Estabrooks indicates, however, such cue-facilitated repeated practice has been shown by most studies to have only short-lived effects, and to have little transfer to noncued situations.

A variety of techniques have been developed based on the approach of guiding patients into novel strategies that utilize residual capacities. One of these, developed by Helm-Estabrooks herself, is Voluntary Control of Involuntary Utterances (VCIU). In this technique, the instructor notes responses that occur as incidental or incorrect utterances. These responses are then used as targets, on the grounds that they are more readily available. This technique has been used with some success to build up a vocabulary of reliably available responses. A related approach is that of "deblocking," first described by Weigl (1968). Weigl found that a word that could not be elicited through one channel (e.g., picture naming) could be primed by first eliciting its production through a reliably functioning route, such as oral reading.

One of the most common bases for failure in name retrieval is the perseverative intrusion of an earlier response. The appearance of such perseverations can be understood in terms of Gordon's discussion of models of naming in Chapter 2, where he describes how such factors as frequency or recency of a response may transiently raise its resting level of activation. Some instances of perseveration (recurrent perseveration) may result from the triggering of a recently given response because its activation level is temporarily higher than that of the target word. In other instances perseveration appears to result from the inability to shift from an immediately preceding response (stuck-in-set perseveration).

Regardless of the mechanism behind the perseverative response, deblocking of the target by preventing perseveration has been used successfully by Helm-Estabrooks and her collaborators with lasting effects. As she points out, the deblocking techniques are not so much based on models of the naming process as on using strategies for benefiting from residual functions. These treatments are contrasted with techniques that attempt to localize each case of anomia to a particular stage in a cognitive model of the naming process. Nettleson and Lesser's (1991) study is prototypical. Using a three-stage model (semantic → phonological lexicon → phonological buffer), they first attempted to diagnose the stage of the damage by a series of tests. Following this they designed a treatment that was intended to be targeted to the damaged process. The effect of model-appropriate treatment was then compared to model-inappropriate treatment. These studies have been done on a small scale with somewhat promising results. However, it is clear that demonstrating the effectiveness of this approach depends on finding selected cases with discrete patterns of success and failure on the screening tests. Helm-Estabrooks remains cautious about the claims of cognitively based approaches to therapy and opts for further exploration of techniques that facilitate performance without directly confronting the defective process.

Difficulties in word finding in general, and object naming in particular, are a common accompaniment of all forms of aphasia. Moreover, such difficulties can remain a serious problem for the individual long after the initial injury. It is largely for these reasons that anomia has remained a central concern from the very beginnings of the systematic study of aphasia. The chapters in this book review the considerable progress that has been made in understanding anomia, at the clinical, cognitive, and anatomic level. Indeed, much of this progress has occurred within the last decade.

Yet there are still gaps in our understanding. One enduring problem is distinguishing between impairments that appear as a general increase in difficulty in vocabulary access in all word-retrieval situations, and those that appear as a complete functional disconnection between semantics and phonological representation. As we have seen, dissociations that are clearly based on injury to an anatomically defined input channel, such as some of the modality-specific anomias, are less puzzling than those that have no known anatomic basis.

The uncertain status of theory-driven therapeutic approaches to the treatment of anomia, as described by Helm-Estabrooks, may well be a manifestation of the gaps that still exist in our understanding of these disorders. The rapid pace of progress in research, however, gives us confidence that the existing gaps in our knowledge will be filled, to the benefit of both the theoretical understanding of, and effective treatment for, anomia.

REFERENCES

Burke, D. M., MacKay, D. G., Worthley, J. S., & Wade, E. (1991). On the tip of the tongue: What causes word finding failures in young and older adults? *Journal of Memory and Language, 30,* 542–579.

Levelt, W. J. M. (1989). *Speaking.* Cambridge, MA: M.I.T. Press.

Geschwind, N. (1969). Problems in the anatomical understanding of aphasia. In A. L. Benton (Ed.), *Contributions to clinical neuropsychology* (pp. 107–128). Chicago: Aldine.

Nettleson, J., & Lesser, R. (1991). Therapy for naming difficulties in aphasia: Application of a cognitive neuropsychological model. *Journal of Neurolinguistics, 6,* 139–157.

Weigl, E. (1968). On the problem of cortical syndromes: Experimental studies. In M. L. Simmel (Ed.), *The reach of mind* (pp. 143–159). New York: Springer.

Author Index

Subject Index

A

Abstract attitude, 196
Acalculia, 3
Acquisition, age of, 36
Action naming, 80–85 (*see also*
    Dissociations; noun–verb)
  in aging, 170–172
  localization, 84
  noun–verb comparisons, 171–173
Action naming test, 170
Aging and name retrieval, 166–176, 207
  cognitive models, 174–176
  confrontation naming, 166–169
  error types in naming, 167
  in discourse, 173–174
  longitudinal studies, 167–168
  naming latencies, 169
  proper nouns, 175
Agnosia
  visual, 93, 98
Agrammatism, 3, 6–7
Agraphia, 3
Alexia, 3
Alzheimer's disease
  access to vocabulary, 5–6
  name retrieval, 166, 176–183, 207
  visual perceptual processes, 93, 177
Amusia, 3
Angular gyrus, 3, 11, 15, 16, 18
Animal naming, *see* Dissociations;
    animate–inanimate)
Anomia
  clinical forms, 6
  definition, 5
  degradation, 201
  storage versus retrieval, 191
Anomic aphasia, *see* Aphasia
Aphasia
  anomic, 9, 16, 18, 96

Broca's, 3, 6–7, 13, 16,18, 95
childhood, 6
conduction, 3, 8, 12, 16, 20
definition of, 3
fluent, 34–35
global, 120
nonoptic, 106
optic, 10, 97, 104, 205
tactile, 97–98
Wernicke's, 3, 7, 12,18, 95
Arcuate fasciculus, 12, 15, 16
Articulation, 13, 41
Attractor dynamics, *see* Models of naming

B

Blends, phonosemantic *see* Error patterns
Body part naming, *see* Dissociations
Boston Naming Test, 167
Broca's aphasia, *see* Aphasia
Broca's area, 4, 5, 16

C

Cascade models, *see* Models of naming
Category-specific dissociations *see*
    Dissociations
Childhood aphasia, *see* Aphasia
Circumlocution, 15
Computational models, *see* Models of
    naming
Concrete entities, 66
Conduction aphasia, *see* Aphasia
Confrontation naming, 10, 31
Connectionist models, *see* Models of
    naming
Consistency in naming
  across time, 180–181
  across tasks, 181–182
Convergence–divergence zones, 66
Cuing, *see* Priming

221

ISBN 0-12-289685-8

90018